MW00676796

Personal Computing

**Communicating
With the
IBM PC Series**

1988   0 471 91667 6

**IBM PC
Upgraders Manual**

1988   0 471 63177 9

**DOS Productivity
Tips & Tricks**

1988   0 471 60895 5

**IBM PS/2
User's
Reference Manual**

1990   0 471 62150 1

Reference

**Data and Computer
Communications:
Terms, Definitions
and Abbreviations**

1989   0 471 92066 5

DIGITAL NETWORKING

DIGITAL NETWORKING
AND T-CARRIER MULTIPLEXING

Gilbert Held

4-Degree Consulting,
Macon, Georgia, USA

JOHN WILEY & SONS
Chichester • New York • Brisbane • Toronto • Singapore

Copyright © 1990 by John Wiley & Sons Ltd.
        Baffins Lane, Chichester
        West Sussex PO19 1UD, England

All rights reserved.

No part of this book may be reproduced by any means,
or transmitted, or translated into a machine language
without the written permission of the publisher.

Other Wiley Editorial Offices

John Wiley & Sons, Inc., 605 Third Avenue,
New York, NY 10158-0012, USA

Jacaranda Wiley Ltd, G.P.O. Box 859, Brisbane,
Queensland 4001, Australia

John Wiley & Sons (Canada) Ltd, 22 Worcester Road,
Rexdale, Ontario M9W 1L1, Canada

John Wiley & Sons (SEA) Pte Ltd, 37 Jalan Pemimpin #05-04,
Block B, Union Industrial Building, Singapore 2057

Library of Congress Cataloging-in-Publication Data

Held, Gilbert
    Digital networking and T-carrier multiplexing / Gilbert Held.
        p.   cm.
    ISBN 0 471 92800 3
    1. Data transmission systems.   2. Computer networks.
    3. Multiplexing.   I. Title.
    TK5105.H427   1990   90-35077
    621.382—dc20   CIP

British Library Cataloguing in Publication Data

Held, Gilbert
    Digital networking and T-carrier multiplexing.
    1. Telecommunication systems. Digital signals. Processing
    I. Title
    621.3822

ISBN 0 471 92800 3

Printed in Great Britain by Courier International Ltd, Tiptree, Essex

CONTENTS

3   Digital Networking Facilities   57

PREFACE

Today we are witnessing a communications revolution whose impact may be as pronounced as the industrial revolution. Instead of machinery producing products for the masses, the key components of the communications revolution are integrated circuits developed to permit voice and data to share common transmission facilities.

Like the industrial revolution, the communications revolution can be expected to provide a wide variety of benefits, foremost of which is the ability to integrate voice and data networks. To provide a common data transportation highway for voice, data, and video transmission, most interexchange carriers (IECs) are rapidly converting their facilities to an all-digital system. In the United States, US Sprint's network was the first to be converted to an all-digital facility in 1989, with AT&T and MCI networks over 95 percent digital at that time. The local exchange carriers (LECs) have kept in step with the IECs, converting most central offices and interoffice circuits to digital facilities. In fact, most analog transmission today occurs over digital communications carrier facilities. With the internal cost of many analog and digital lines virtually identical, it is only a matter of time until communications carriers do not differentiate in price between the two. As this occurs, the higher transmission quality and data rate obtained from the use of digital transmission facilities will result in a quantum leap in the use of those facilities. With this in mind, this book was written to provide readers with an understanding of digital transmission technology and how this technology will provide the data transportation highway of the future.

In this book we will focus our attention upon digital networking to include the advantages and disadvantages associated with digital signaling, the operation and utilization of communications carrier offerings, the frame formats of high-speed North American and

European T-carriers, the features and operational utilization of T-carrier multiplexers, and the methods by which digital facilities can be tested. Each of the previously mentioned areas is investigated in a hierarchical manner, with each chapter building upon the material presented in the preceding chapter. It is, therefore, recommended that readers unfamiliar with digital networks and digital signaling techniques should read each chapter in sequence. More experienced readers requiring specific information, such as testing and troubleshooting methods, can consider beginning their reading at a specific chapter in this book.

In developing the material incorporated into this book, I have structured it as a classroom text for students that previously completed an introductory course in data communications. In addition, this book was also written for the data communications practitioner working in industry or government that has equivalent experience through on-the-job training and the reading of appropriate introductory books, college courses, or seminar attendance. For both categories of readers, without appearing to be self-serving, I would recommend reading *Understanding Data Communications: From Fundamentals Through Networking*, which I previously authored, to provide readers with a comprehensive introduction to the field of data communications.

Since this book was primarily written as an advanced text for students completing an introductory course in data communications, it should be considered as a one-semester course for use by seniors or first-year graduate students majoring in telecommunications or computer science. To assist both students and practitioners, I have included a series of review questions at the end of each chapter. These questions were selected to reinforce key concepts and readers are encouraged to work each problem prior to going forward in the book.

As a professional author, I sincerely welcome reader comments. Thus, I encourage you to write to me through my publisher or directly to obtain feedback concerning the material presented in this book.

Gilbert Held
Macon, Georgia

ACKNOWLEDGEMENTS

Similar to several of my previously published books, a portion of the material incorporated into this book was originally developed for a series of high-technology seminars presented in the United States and Europe. Once again, I wish to thank Mr Joseph Savino of Frost & Sullivan, Inc., for providing me with the opportunity to develop and teach a digital networking seminar in Europe which provided the impetus for writing this book.

Like fine music produced by an orchestra, the publication of a book is a team effort. Thus, I would like to thank Mr Ian McIntosh and Ms Ann-Marie Franks of John Wiley & Sons Ltd for their cooperation and assistance in moving the manuscript through its assorted channels. I would also like to thank Mrs Carol Ferrell for once again taking my notes and diagrams and turning them into a manuscript suitable for publication. Last, but not least, I must also thank my family for providing the time and having patience while I worked on this book.

1

RATIONALE FOR DIGITAL NETWORKING

In this chapter we will examine why both large and small business organizations, communications carriers, government agencies, and universities are migrating their voice and data communications to digital networks. Although there are many key advantages obtained from digital networking, as might be expected from the use of any technology, there are also disadvantages associated with transmission occurring via a digital network.

To obtain an understanding of the advantages and disadvantages of digital networking first requires a comparison of the method of operation of analog and digital transmission systems. Once this comparison is completed we will use this information as a base to explore the advantages and disadvantages of digital signaling which forms the basis for the construction of digital networks. This information will provide us with the rationale behind the large investments both communications carriers and end-user organizations are making to obtain a digital networking capability.

1.1 ANALOG TRANSMISSION

In analog transmission a continuous wave is modulated to impress information onto a carrier signal. The resulting modulated signal is subject to several types of impairments as it propagates down a transmission path, including attenuation and envelope delay.

The resistance, capacitance, and inductance of a circuit attenuate the signal, resulting in a reduction in its signal strength.

In addition, the construction of communications carrier analog voice-grade facilities is such that only frequencies between 300 Hz and 3300 Hz are passed, forming the voice-grade passband illustrated in Figure 1.1. This passband, which is a contiguous portion of an area in the frequency spectrum, is formed by the use of low-pass and high-pass filters and results in a bending of the attenuation-frequency response of an analog signal as it propagates along the transmission path. Since high frequencies attenuate more rapidly than low frequencies, this natural phenomenon results in a further distortion of the received signal.

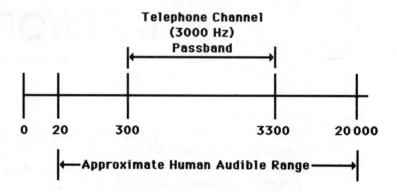

Figure 1.1 Passband of a telephone channel. The telephone channel passband is a contiguous area of the frequency spectrum formed by the use of filters which permits frequencies between 300 and 3300 Hz to pass

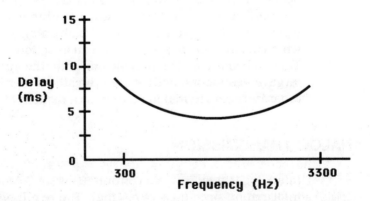

Figure 1.2 Envelope delay curve. Both the filters used by a communications carrier and the difference in the velocity of the harmonics of a complex signal result in some portions of a signal arriving at its destination prior to other parts of the signal. This delay is known as envelope delay

A second problem associated with analog transmission is caused by the delay of certain frequencies of a composite signal as they flow through a communications channel. Usually, the filters employed by the communications carrier delay frequencies near the cutoff frequency of the filters. In addition, the non-linear relationship between the phase shift and frequency of analog signals results in the propagation of the harmonics of the signal at different velocities. Together, the filters used by the communications carriers and the difference in the velocity of propagation of the harmonics of a complex signal result in an envelope delay curve similar to that illustrated in Figure 1.2.

1.1.1 Amplifiers and equalizers

At periodic intervals throughout their network, communications carriers have installed amplifiers to compensate for the loss in strength of analog signals. Although an amplifier will boost signal strength, it also increases any noise or distortion that previously occurred to the signal. This is illustrated in Figure 1.3. Thus, after amplification, the effect of noise and distortion that may have previously occurred is also increased.

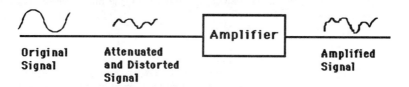

**Original
Signal**   **Attenuated
and Distorted
Signal**   **Amplifier**   **Amplified
Signal**

Figure 1.3 Amplification of an analog signal. Although an amplifier boosts the strength of an analog signal, it also increases the effect of any noise and distortion that occurred to the signal

To minimize the effect of the amplitude–frequency response of a signal and any envelope distortion, analog leased lines can be conditioned. What is known as C-level conditioning in the United States results in the communications carrier installing equalizers to generate both an amplitude–frequency response and a delay inverse to that occurring to a signal traversing the circuit. The resulting objective of the use of equalizers is to produce a uniform delay and uniform amplitude–frequency response over the passband.

Figure 1.4 illustrates the theoretical effect obtained by the use of attenuation equalizers, while Figure 1.5 illustrates the theoretical effect obtained by the use of delay equalizers. As illustrated in Figure 1.4, the attenuation equalizer introduces a variable gain inverse to the attenuation loss occurring on the circuit, theoretically resulting in a uniform attenuation across the frequency spectrum. Similarly, the use of delay equalizers is designed to provide a uniform delay to all components of a signal over the voice channel passband. This is accomplished by the delay

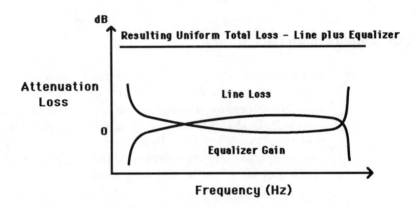

Figure 1.4 Using an attenuation equalizer. An attenuation equalizer introduces a variable gain inverse to the line loss in an attempt to obtain a uniform total loss

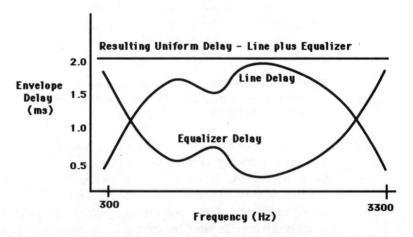

Figure 1.5 Using a delay equalizer. An attenuation equalizer introduces a variable delay inverse to the line delay in an attempt to obtain a uniform total delay

equalizer introducing a variable delay inverse to the actual delay occurring on the circuit.

In actuality, the equalizers used by communications carriers are fixed and are set at circuit installation time. Since line conditions vary, the equalizers are used to provide a range of acceptable values that the carrier attempts to provide, based upon the level of conditioning selected. Tables 1.1 and 1.2 list the bandwidth parameter limits for frequency loss in decibels (dB) and envelope distortion delay in microseconds (μs), respectively. The BASIC entry references an unconditioned channel, while C1, C2, C4, and C5 represent four types of conditioning available from AT&T. British Telecom Keyline bandwidth parameter limits are the same as the Consultative Committee for International Telephone and Telegraph (CCITT) M.1020 four-wire international leased line recommendation. Note that a negative dB limit in Table 1.1 means less loss or a gain while a positive dB limit is more loss or an actual signal loss.

For high-speed transmission on the switched telephone network

Table 1.1 Bandwidth parameter limits—frequency response loss.

Channel/conditioning	Frequency range	Limits in dB
BASIC	300–500	−3 to +12
	500–2500	−2 to +8
	2500–3000	−3 to +12
C1	300–1000	−3 to +12
	1000–2400	−1 to +3
	2400–2700	−2 to +6
	2700–3000	−3 to +12
C2	300–500	−2 to +6
	500–2800	−1 to +3
	2800–3000	−2 to +6
C4	300–500	−2 to +6
	500–3000	−2 to +3
	3000–3200	−2 to +6
C5	300–500	−1 to +3
	500–2800	−0.5 to +1.5
	2800–3000	−1 to +3
British Telecom Keyline (CCITT M.1020)	300–500	−2 to +6
	500–2800	−1 to +3
	2800–3000	−2 to +6

Table 1.2 Bandwidth parameter limits—envelope distortion delay.

Channel/conditioning	Frequency range	Limits (μs)
BASIC	800–2600	1750
C1	800–1000	1750
	1000–2400	1000
	2400–2600	1750
C2	500–600	3000
	600–1000	1500
	1000–2600	500
	2600–2800	3000
C4	500–600	3000
	600–800	1500
	800–1000	500
	1000–2600	300
	2600–2800	500
	2800–3000	3000
C5	500–600	600
	600–1000	300
	1000–2600	100
	2600–2800	600
British Telecom Keyline (CCITT M.1020)	500–600	3000
	600–1000	1500
	1000–2600	500
	2600–3000	3000

and very high-speed transmission on leased lines, modem manufacturers have included automatic and adaptive equalization circuitry in their products. Each time the direction of data flow changes, one modem sends a 'training' signal which is used by the other modem to adjust its level of equalization. This process of automatic and adaptive equalization affects data throughput as well as adds to the complexity and expense of the modem.

Another problem associated with the use of automatic and adaptive equalizers is the training time that can be required after a line hit. Some modems can take as long as several seconds to retrain after a line hit, resulting in the absence of data being transmitted during that time. If a time-dependent protocol is being used when a line hit occurs, the modem retraining time can result in a timeout, causing the protocol to go into a reinitialization routine which further reduces transmission throughput.

1.2 DIGITAL TRANSMISSION

When digital transmission facilities are used, data travels from end-to-end in a digital format. Although digital pulses are subject to impairments similar to those experienced by analog signals, carrier facilities employ repeaters instead of amplifiers which significantly reduces the effect of distortion upon the signal.

1.2.1 Use of repeaters

At regular intervals on a digital transmission facility, the communications carrier installs repeaters whose function is to rebuild or regenerate pulses into their original form.

Figure 1.6 illustrates the operational effect from the use of a repeater in a digital transmission system. The repeater scans the line looking for a pulse rise and then discards the rest of the pulse, regenerating a new digital pulse in place of the incoming pulse. This operation results in the elimination of any prior distortion and explains why, in general, a digital transmission system is more reliable than an analog transmission system that employs amplifiers which boost both the signal and any prior distortion the signal received.

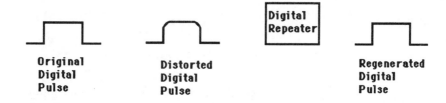

Original Digital Pulse   Distorted Digital Pulse   Digital Repeater   Regenerated Digital Pulse

Figure 1.6 Repeaters eliminate distortion. When digital transmission facilities are used, data travels from end-to-end in a digital format. Digital pulses are regenerated at regular intervals which results in the elimination of distortion

1.3 ADVANTAGES AND DISADVANTAGES OF DIGITAL SIGNALING

Digital networks are based upon the use of digital signaling. Thus, we can examine the basis for the growth in digital networking by examining the advantages and disadvantages associated with digital signaling.

Table 1.3 lists the primary advantages and disadvantages of digital signaling in comparison to analog signaling. By reviewing each of the entries in the table we can obtain an appreciation behind the rationale for the growth in the use of digital transmission facilities as well as some of the problems that can be expected to occur when using this technology.

Table 1.3   Advantages and disadvantages of digital sampling.

Advantages	Reliability
	Multiplexing efficiency
	Voice and data integration capability
	System performance monitoring capability
	Ease of encryption
	Economics
	Path to ISDN
Disadvantages	Loss of precision in analog to digital conversion
	Analog system interface requirements
	Increased bandwidth requirements
	Need for precise timing

1.3.1 Advantages

Reliability

In terms of reliability, a digital network can be expected to have a lower error rate than an analog network. This expectation is based upon the previously discussed differences between amplifiers and repeaters with respect to an analog signal being boosted to include any noise and distortion, while a digital pulse is discarded by the repeater which then generates a completely new pulse. A second significant difference between analog and digital signaling reliability results from the equalizers required for high-speed analog transmission. Although equalizers are built into digital service units (DSUs) which are the digital equivalent of a modem, their circuitry is far less complex than equivalent circuitry incorporated into modems. In addition to DSUs having a cost significantly less than that of a high speed modem, their lower complexity makes them more reliable. Finally, the retraining time on a digital network is usually significantly less than the retaining time of modems used on analog networks, in effect significantly reducing the possibility of a timeout occurring.

Multiplexing efficiency

When comparing the ease of multiplexing analog to digital signals, we must compare the technology associated with each technique. In frequency division multiplexing (FDM), tuned filters are required to derive subchannels from the available frequency spectrum of a circuit. To prevent frequency drift resulting in the modulated data on one subchannel interfering with the transmission of data on adjacent subchannels, guard bands are assigned between subchannels. Figure 1.7 illustrates the splitting of a 3 kHz voice channel into subchannels, with guard bands employed to minimize the effect of frequency drift between channels. The guard bands are unused segments of frequency which alleviate the effect of frequency drift associated with analog devices. Unfortunately, the use of guard bands also removes portions of the frequency spectrum that could be used for transmission, reducing the overall efficiency of frequency division multiplexing. A second limitation associated with FDM equipment is the channel spacings required for multiplexing different data rates. As indicated in Table 1.4, FDM channel spacing requirements would not enable two 1200 bps data sources to be multiplexed on a conventional voice-grade channel since only 3000 Hz is available for use and each 1200 bps source would require 1800 Hz.

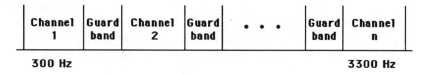

Frequency

Figure 1.7 FDM channel separations. In frequency division multiplexing, the 3 kHz bandwidth of a voice-grade line is split into channels or data bands separated from each other by guard bands

Table 1.4 FDM channel spacings.

Speed (bps)	Spacing (Hz)
75	120
110	170
150	240
300	480
450	720
600	960
1200	1800

The use of time division multiplexers (TDM) to support digital transmission can result in a higher level of efficiency. This is because the aggregate data transmission capacity of the composite channel is the limiting constraint and the multiplexer interleaves data from low to medium speed sources onto the high-speed channel by time.

In comparing FDM to TDM one must also consider the reliability of equipment and its ease of modification. In both cases, TDM equipment holds an advantage over FDM equipment since digital components are more reliable than analog components and it is easier to adjust the assignment of time slots than vary subchannel separations that are controlled by the use of tuned filters.

Voice and data integration capability

Since the early 1960s, communications carriers have been adding equipment and modifying their network facilities to support digital signaling. At carrier central offices, equipment has been added that digitizes voice conversations, enabling both voice and data to be multiplexed and transmitted together on trunk circuits connecting carrier offices.

The success experienced by communications carriers in the integration of voice and data eventually resulted in the commercial offering of high-speed communications facilities known as T-carrier circuits. This, in turn, resulted in equipment vendors developing products that enabled end-user organizations to design their own integrated voice/data networks using T-carrier facilities. Both the operation and utilization of T-carrier facilities, as well as the use of multiplexers containing different types of voice digitization cards, are described later in this book.

System performance monitoring capability

In an analog transmission system the measurement of circuit performance requires the insertion of a known pattern of test bits which interrupts the flow of data. In comparison, two framing formats used for T-carrier transmission include a mechanism whereby performance data can be obtained by monitoring the circuit without interrupting the flow of data. A second advantage of digital signaling with respect to analog signaling is the coding mechanism used to place digital data onto a digital transmission facility. This coding mechanism results in the ability of equipment to identify errors without requiring the equipment to have knowledge of the information that is being transmitted. These errors are detected as bipolar violations (BPV), and both digital encoding techniques and bipolar violations are discussed in detail later in this book.

Ease of encryption

Although analog transmission can be encrypted, to do so is both difficult and costly. In addition, its result may be reversible within a short period of time. This is because the encryption of an analog signal requires the use of numerous filters to divide the frequency spectrum into subchannels. The resulting subchannels are then moved to different positions within the frequency spectrum, in effect making a telephone conversation indecipherable to a person monitoring the transmission.

When digital signaling is used, encryption is both more secure as well as easier to implement. This is because a pseudo-random bit string can be generated by a key and added via modulo-2 addition to the source digital data stream to encrypt transmission. Then the ability to change the key used to generate the pseudo-random bit string allows almost an infinite number of codes to be generated. In comparison, the bandwidth of an analog voice-grade line and the physical constraints in developing filters and circuitry necessary to scramble a voice conversation results in a relatively low number of channel positioning possibilities in comparison to the number of pseudo-random bit strings that can be generated using digital circuitry.

Economics

On a cost per bit per second (bps) basis, digital transmission facilities are in certain situations more economical than analog facilities.

The maximum data rate obtainable on analog leased lines was 19.2 kbps using trellis coded modulation modems that were very expensive when this book was written. Since then one vendor has introduced a leased line modem capable of operating at 24.4 kbps. In comparison, digital transmission facilities can be obtained to provide a transmission rate of 2.4 to 56 or 64 kbps for the use of subrate facilities and 1.544/2.048 Mbps on T-carriers by the use of inexpensive digital service units. Here 1.544 Mbps is the T-carrier data transmission rate in North America while the 2.048 Mbps data rate is used for T-carrier transmission in Europe, which is also known as E1 transmission.

Although there are numerous factors that contribute to the precise rate paid for analog and digital circuits to include access connection fees, variable interoffice channel charges based upon the distance between locations to be connected, and installation cost, for simplicity we will focus our attention upon mileage cost. For illustrative purposes, Table 1.5 lists the approximate costs per mile for an analog leased line and several types of digital transmission facilities for distances between locations exceeding 500 miles that were in effect by one communications carrier when this book was written. Obviously, for an up-to-date economic comparison, the reader should contact one or more communications carriers to obtain both the current interoffice channel cost and other fees appropriate to each type of circuit under consideration.

Table 1.5 Interoffice channel cost, cost per mile, distances > 500 miles.

Facility	Cost per mile (¢)
Analog circuit	0.32
2.4 kbps	0.32
4.8 kbps	0.32
9.6 kbps	0.32
56.0 kbps	1.69
1.544 Mbps T-carrier	15.50

For the cost effectiveness examples that follow we will assume that the cost of a 19.2 kbps modem is $4000 per unit. In addition, let us assume that the cost of a DSU operating at data rates up to 9.6 kbps is $1368 while the cost of a 56 kbps DSU is $1444. For a T-carrier circuit operating at 1.544 Mbps, the DSU function is normally built into multiplexing equipment, and another device known as a channel service unit (CSU) is used for transmission on that type of circuit. Thus, we will assume that the cost of a T-carrier CSU is $1750 in the economic comparisons we will perform. Finally, in each of the following economic comparisons we will assume that purchased equipment, such as modems or DSUs and CSUs, are amortized over a three-year (36-month) period.

Thus, the cost of a 500-mile analog circuit operating at 19.2 kbps on a per bps per month basis becomes

$$\frac{(2 * \$4000/36 \text{ months}) + 0.32/\text{mile} * 500 \text{ miles}}{19.2 \text{ kbps}}$$
$$= 0.0199 \text{ per bps per month}$$

Now let us examine the cost of several types of digital transmission facilities. The cost of a 500-mile digital facility operating at 2.4 kbps on a per bps per month basis is

$$\frac{(2 * \$1368/36 \text{ months}) + 0.32/\text{mile} * 500 \text{ miles}}{2.4 \text{ kbps}}$$

$$= 0.098 \text{ per bps per month}$$

Using the same method of cost computation, the monthly cost per bps for a 500-mile digital circuit operating at 4.8 kbps becomes

$$\frac{(2 * \$1368/36 \text{ months}) + 0.32/\text{mile} * 500 \text{ miles}}{4.8 \text{ kbps}}$$

$$= 0.049 \text{ per bps per month}$$

Now let us compute the cost of a 500-mile digital circuit operating at 9.6 kbps on a per bps per month basis. This cost is computed as follows:

$$\frac{(2 * \$1368/36 \text{ months}) + 0.32/\text{mile} * 500 \text{ miles}}{9.6 \text{ kbps}}$$

$$= 0.0245 \text{ per bps per month}$$

Note from the preceding analysis that the cost of low-speed digital transmission facilities is normally more expensive than an analog circuit on a per bit per month basis. While there are many other factors that should be investigated, a digital transmission facility does not always imply an economic saving over an analog facility. Now let us examine the cost of two additional digital facilities. The cost of a 500-mile 56 kbps digital facility on a per bps per month basis is

$$\frac{(2 * \$1444/36 \text{ months}) + 1.69/\text{mile} * 500 \text{ miles}}{56.0 \text{ kbps}}$$

$$= 0.0165 \text{ per bps per month}$$

The cost of a 500-mile 1.544 Mbps T-carrier circuit on a per bps per month basis is

$$\frac{(2 * \$1750/36 \text{ months}) + 15.50/\text{mile} * 500 \text{ miles}}{1.544 \text{ Mbps}}$$

$$= 0.005 \text{ per bps per month}$$

In Table 1.6, the reader will find a summary of the previously computed costs of the voice-grade analog circuit and the five digital circuits on a per bps per month basis. Note that only when a digital transmission facility operates at or above 56 kbps is the cost per bps per month lower than on an analog transmission facility. In fact, the cost of a 1.544 kbps T-carrier digital facility is approximately one-quarter that of an analog voice-grade circuit on a bps per month basis.

Table 1.6 Cost comparisons: analog vs. digital circuits.

Type of circuit	Cost per bps per month*
Analog voice-grade circuit	
19.2 kbps	0.0199
Digital circuits	
2.4 kbps	0.098
4.8 kbps	0.049
9.6 kbps	0.0245
56.0 kbps	0.0165
1.544 Mbps	0.005

* Cost computed based upon interoffice channel costs contained in Table 1.5 and equipment amortized over 36 months.

A second method of comparing the cost of analog and digital transmission facilities involves examining the voice and data carrying capacity of a T-carrier circuit. In North America, a T-carrier can be used to transmit data, a mixture of digitized voice and data, or it can be used to transmit at least 24 digitized voice conversations. If we assume the latter, the economics of using a T-carrier facility becomes substantially more pronounced. As an example of this, in early 1990 the cost of an analog voice-grade circuit between Macon, Georgia, and Washington, DC—a distance of about 650 miles—was approximately $900 per month. In comparison, a T-carrier circuit capable of supporting 24 digitized voice circuits cost approximately $8000 per month. Disregarding the cost of multiplexing equipment, approximately nine (8000/900) analog circuits would justify the use of a T-carrier facility that could support 24 voice circuits.

Another method of comparing analog and digital transmission facilities is on an aggregate transmission capacity basis. The highest data transmission rate obtainable on an analog circuit using commonly available modems is 19.2 kbps, although one vendor announced a higher-speed modem operating at 24.4 kbps when this book was published. When a 19.2 kbps data rate is compared to the 1.544 Mbps data transmission rate of a T-carrier facility, that facility can carry the equivalent of approximately 80 analog data channels even though its cost may be less than one-tenth that number of analog circuits.

Path to ISDN

ISDN is defined by the CCITT as 'a network, evolved from the telephone network, that provides end-to-end digital connectivity to support a wide range of services, including voice and non-voice services, to which users have access by a limited set of standard multipurpose customer interfaces'. ISDN services fall into three general categories: basic rate, primary rate, and broadband. The latter category of services is not expected to be commercially available until the turn of the century.

The basic rate interface consists of two 64 kbps channels that can be used to carry voice or data and a 16 kbps channel that carries supervisory and signaling information. Each 64 kbps channel is known as a bearer channel, giving rise to the term 'B channel'. The supervisory and signaling channel is called a D channel. These three channels, which are designed to provide a subscriber interface from a private branch exchange (PBX) to the end-user, are carried on a single twisted pair wire at a composite data rate of 144 kbps and are referred to as 2B + D.

The primary rate interface is designed to provide an interconnection between ISDN PBXs and the ISDN network. The data transmission rate of the primary rate interface is based upon the use of current T-carrier facilities. Although T-carrier facilities and the ISDN primary rate allocate channels differently, both operate at a rate of 1.544 Mbps in North America and 2.048 Mbps in countries that follow the CCITT standard.

When the basic frame format used on a North American T-carrier circuit is examined, it is seen that it consists of 24 eight-bit time slots plus a framing bit, resulting in a frame containing 193 bits. The frame frequency is 8000 frames per second, resulting in each time slot having a 64 kbps capacity, as illustrated in Figure 1.8A. When voice is carried in a time slot, one bit in every 48 bits may be 'robbed' for signaling, resulting in the data transmission rate being limited to 56 kbps. In Europe, the T-carrier circuit consists of thirty 64 kbps time slots used for data and/or voice, as well as a time slot used for signaling, and a time slot used for synchronization. This is illustrated in Figure 1.8B. In North America, the ISDN primary rate interface consists of 23 B channels used to carry voice or data and

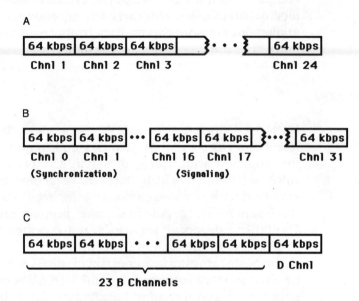

Figure 1.8   Framing comparisons: A North American T-carrier frame format; B European T-carrier frame format; C North American primary rate ISDN frame format

a D channel that carries all signaling information as illustrated in Figure 1.8C. In Europe, a 30B + D PRI format is used for ISDN, with all signaling information moved to the D channel. Thus, a T-carrier circuit provides the stepping stone from the high-speed networks of today to ISDN primary rate interface services of tomorrow as long as the equipment used can correspond to ISDN PRI framing requirements.

1.3.2 Disadvantages

Now that we have examined the advantages associated with digital signaling, let us focus our attention upon some of the problems associated with this technology.

Loss of precision

Transmitting a voice conversation on a digital network requires the conversion of analog signals to digital signals. Since an analog signal is continuous, it can take on an infinite number of values. In comparison, digital signals represent discrete values. Thus, the conversion from analog to digital can result in a loss of precision. This loss, known as quantizing noise, can result in a distortion of a voice conversation when the digital signals are reconverted into their analog form. In Chapter 2, the reader will find additional information concerning quantizing noise.

Analog system interface

Prior to the use of digital signaling, ringing, on-hook and off-hook status, as well as dialed digit, information was conveyed by the use of relatively high current and high voltage. Digital circuitry cannot tolerate high current or voltage, resulting in the necessity of adding protective circuitry to digital devices. Although protective circuitry is most effective in eliminating most high current and voltage problems, it slightly raises the cost of interfacing digital devices to existing analog facilities.

Increased bandwidth requirements

In 1924, Nyquist showed that the maximum signaling rate on a channel expressed in baud (B) was related to the bandwidth (W) expressed in Hz as follows:

$$B = 2W$$

The preceding limitation avoids intersymbol interference, a condition in which one transmitted signal interferes with a succeeding signal.

Although the passband of an analog telephone channel is 3 kHz, in actuality, approximately 4 kHz are used due to the bending of the attenuation-frequency response at the cutoff filter frequencies and the separation of one channel from another when frequency division multiplexing is used by a communications carrier to combine several analog signals onto a trunk routed between carrier offices. Thus, the support of 24 channels in an analog carrier system would require 4 kHz×24 or 96 kHz.

The signal rate of 1.544 Mbps for a North American T-carrier requires a bandwidth of 772 kHz, based upon the Nyquist relationship between signal rate and bandwidth. Thus, a T-carrier supporting 24 voice channels requires eight times the bandwidth of an analog carrier system.

Need for precise timing

Digital signaling techniques employ synchronous transmission. Since synchronous systems require the detection of the presence of bits by sampling at precise times, timing or clocking information is required. Not only does this raise the cost of digital signaling, but, in addition, results in a variety of errors when timing information becomes distorted or lost.

Cost

As noted from the previously performed economic comparisons, certain digital transmission facilities may be more costly than analog facilities. In general, low-speed digital transmission facilities will cost more than analog circuits used for data transmission, while a T-carrier circuit capable of supporting 24 or 30 voice channels can be economically justified by a requirement for less than half that number of channels.

REVIEW QUESTIONS

1 Why is it possible for an amplifier to create an error condition?

2 How can an automatic and adaptive equalizer cause a protocol timeout?

3 What is the primary advantage associated with the use of repeaters instead of amplifiers?

4 What is the purpose of a guard band, and why does it decrease the efficiency of a frequency division multiplexer?

5 Discuss five key advantages of a digital transmission system in comparison to an analog transmission system.

6 Can you assume that digital transmission facilities are more economical than analog transmission facilities? Why?

7 Explain why the conversion of an analog signal into a digital signal can result in a loss of precision.

8 What would be the difference in bandwidth requirements between 32 analog voice channels and a T-carrier digital transmission facility operating at 2.048 Mbps?

2

FUNDAMENTALS
OF DIGITAL
NETWORKING

In this chapter, we will examine four interrelated topics to obtain
a firm understanding of the characteristics and operation of digital
networks. To understand why digital signals are modified prior
to transmission on a digital facility, we will first examine the
characteristics of several types of digital signals. This examination
will provide the reader with an understanding of the selection of
the encoding techniques used on communications carrier digital
facilities and why user equipment must be compatible with that
encoding technique.

Since the major advantage obtained from digital networking is the
ability to integrate voice and data, we will next examine a variety of
analog to digital conversion techniques. This will be followed by an
examination of telephone company analog and digital transmission
facilities and a discussion of the evolution of the T-carrier which
forms the basis for the construction of digital networks.

2.1 DIGITAL SIGNALING CHARACTERISTICS

One of the most critical issues to be addressed in the design of
digital transmission facilities is the method by which binary data
will be encoded as signal elements for transmission. The selection
of one encoding method over another affects both the cost of
constructing transmission facilities as well as the resulting quality
of transmission obtained from the use of such facilities. To obtain

an understanding of the advantages and disadvantages of different types of digital signals, let us examine the characteristics of seven types of digital encoding techniques.

2.1.1 Unipolar non-return to zero

Unipolar non-return to zero is a simple type of digital signaling which was originally used for early telegraphy. Today, unipolar non-return to zero signaling is used with private line teleprinter systems, as well as the signal pattern used by the RS-232/V.24 interface.

In this signal scheme, a dc current or voltage represents a mark or binary one, while the absence of current or voltage represents a space or binary 0. This signal encoding technique, which is illustrated in Figure 2.1, is called non-return to zero because the current or voltage does not return to zero between adjacent one bits. When used with a transmission system, line sampling determines the presence or absence of current or voltage, which is translated into an equivalent mark or space.

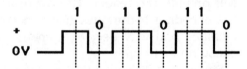

Figure 2.1 Unipolar non-return to zero signaling. In unipolar non-return to zero signaling, the current or voltage does not return to a zero value between adjacent 'one' bits

Although it is possible to use unipolar non-return to zero signaling in low-speed transmission systems, the high data rates of most digital networks make this encoding technique undesirable. Most of the unsuitability of unipolar non-return to zero signaling is due to this signal method having a residual dc component. The residual dc component causes a direct physical attachment of transmission components in network construction. In comparison, a signal encoding method that has no dc component permits coupling to occur via the use of transformers which provide electrical isolation as well as reduce interference.

2.1.2 Unipolar return to zero

Unipolar return to zero is a variation of unipolar non-return to zero signaling. Here the signal always returns to zero after every one bit. While this signal is easier to sample, as each mark has a pulse rise, it requires more circuitry to implement and is not commonly used. In addition, similar to unipolar non-return to zero signaling, unipolar return to zero results in dc voltage buildup. Figure 2.2 illustrates an example of unipolar return to zero signaling.

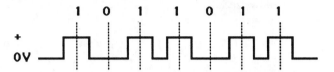

Figure 2.2 Unipolar return to zero signaling. In unipolar return to zero signaling, the signal always returns to a zero level after each 'one' bit

2.1.3 Polar non-return to zero

In polar non-return to zero signaling a positive current is used to represent a mark and a negative current is used to represent a space. This signaling technique eliminates some of the residual dc buildup associated with unipolar signaling, since over a long period of time the number of binary zeros and binary ones will be equal. As no transition occurs between two consecutive bits of the same value, the signal must be sampled to determine the value of each received bit. Figure 2.3 illustrates an example of polar non-return to zero sampling.

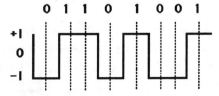

Figure 2.3 Polar non-return to zero signaling. In polar non-return to zero signaling, positive and negative currents are used to represent binary 'ones' and 'zeros', respectively Since no transition occurs between successive 'ones' and 'zeros', line sampling must be used to determine the value of each bit

2.1.4 Polar return to zero

The polar return to zero signal is similar to the polar non-return to zero signal in that it uses opposite polarities of current. However, this signal returns to zero after each bit is transmitted. Figure 2.4 illustrates an example of polar return to zero signaling. Note that a sequence of zeros or a sequence of ones can result in a dc voltage buildup. In addition, since there is a pulse that has a discrete value for each bit, sampling of the signal is not required which reduces the circuitry required to determine whether a mark or space has occurred.

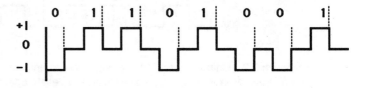

Figure 2.4 Polar return to zero signaling. In polar return to zero signaling, opposite polarities of current are used to represent binary 'ones' and 'zeros', with a return to zero current level after each bit. By returning the current value to zero after each bit, line sampling is not required to determine each bit value

2.1.5 Bipolar non-return to zero

Two signals similar to the return to zero encoding method, but which eliminate the problem of dc voltage component buildup, are bipolar non-return to zero and bipolar return to zero signaling. In bipolar non-return to zero signaling alternating polarity pulses are used to represent marks, while a zero pulse is used to represent a space. Figure 2.5 illustrates an example of bipolar non-return to zero signaling. Note that this encoding method does not require line sampling since the voltage levels can be examined to determine the state of the signal.

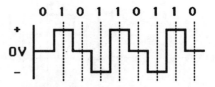

Figure 2.5 Bipolar non-return to zero signaling. In bipolar non-return to zero signaling, binary 'ones' are represented by alternating voltage polarities, while binary 'zeros' are represented by a zero voltage level

2.1.6 Bipolar return to zero

In bipolar return to zero signaling, the bipolar signal returns to zero after each mark, as illustrated in Figure 2.6. This type of signaling insures that there is no dc voltage buildup on the line, which results in ac coupling being accomplished by the use of transformers to provide electrical isolation and reduce the possibility of interference occurring when power and data are carried on a common line. In addition, this method of signal encoding permits repeaters to be placed relatively far apart in comparison to other signaling techniques. This signaling technique is employed in modified form on digital networks due to the economics associated with spacing repeaters further apart from one another.

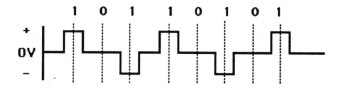

Figure 2.6 Bipolar return to zero signaling. In bipolar return to zero signaling, alternate polarities are used to represent binary 'ones' with the voltage returning to zero after each 'one'

2.1.7 Fifty percent bipolar waveshape

To assist in eliminating high-frequency components that can interfere with other transmissions, digital services utilize 50 per-

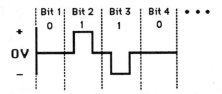

Figure 2.7 Bipolar (AMI) RTZ 50 percent duty cycle. By concentrating the transmitted power in the middle of the transmission bandwidth, AMI signaling minimizes the distortion that may occur to a signal while eliminating dc voltage buildup on the line

cent bipolar waveshape or duty cycle. This signaling technique, which is illustrated in Figure 2.7, concentrates the transmitted power in the middle of the transmission bandwidth. Since the transmission quality of a channel is worse near the edges of the channel, this encoding technique minimizes the distortion that may occur to the resulting signal. The resulting bipolar pulse is also known as alternate mark inversion or AMI signaling.

2.1.8 Bipolar violations

Bipolar transmission requires that each data pulse representing a logical one is transmitted with alternating polarity. A violation of this rule is defined as two successive pulses that have the same polarity and are separated by a zero level.

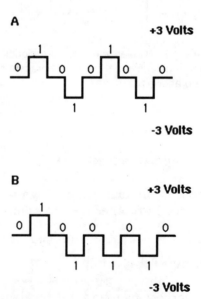

Figure 2.8 Bipolar violations. Two successive negative or positive pulses represent a bipolar violation of a bipolar return to zero signaling technique: A bipolar coding of data; B bipolar violation

A bipolar violation indicates that a bit is missing or miscoded. Some bipolar violations are intentional and are included to replace a long string of zeros that could cause a loss of timing and receiver synchronization or to transmit control information. Figure 2.8 illustrates one example of a bipolar violation. Figure 2.8A shows the correct encoding of the bit sequence 010101010 using bipolar return to zero signaling. In Figure 2.8B, the third one bit is encoded as a negative pulse and represents a violation of the bipolar return to zero signaling technique where ones are alternately encoded as positive and negative voltages for defined periods. In Chapter 4, we will examine several methods used to develop bipolar violations that are used to maintain synchronization when a string of consecutive zeros is encountered. These methods are commonly referred to as zero suppression codes.

2.2 ANALOG-TO-DIGITAL CONVERSION TECHNIQUES

One of the major benefits associated with the use of high-speed digital networks is the ability they afford the user to integrate both voice and data transmission requirements onto a common transmission facility. In this section, we will examine several methods that are commonly used to digitize voice, and the advantages and disadvantages associated with each method.

Analog transmission requires the use of a continuously varying signal. In comparison, digital transmission requires the use of discrete signal levels. Thus, each of the techniques discussed in this section will result in a loss of precision, since the analog signal that can represent an infinite number of values must be converted into a discrete digital signal and then be reconverted into its equivalent analog form.

2.2.1 Pulse code modulation (PCM)

The earliest method used by communications carriers to convert the continuously varying subscriber voice (analog) signal into a digital data stream was PCM. PCM is still the most common technique utilized to digitize voice due to the prior investment of communications carriers in that technology. This technique requires three processing steps—sampling, quantization, and coding.

Sampling

During the PCM sampling step, the analog signal is sampled 8000 times per second. This sampling rate is based upon the Nyquist theorem which states that to truly represent a signal the sampling rate must be twice the highest frequency of the signal to be transmitted. Due to a voice bandwidth being approximately 4000 Hz, the sampling rate is 8000 times per second or every 125 microseconds.

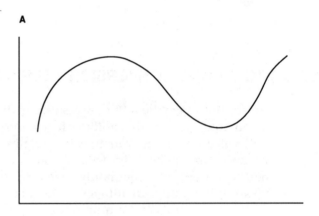

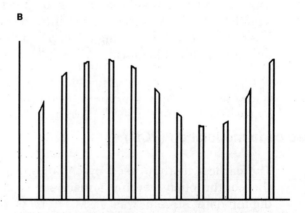

Figure 2.9 Pulse amplitude modulation (PAM). A pulse aplitude modulation (PAM) signal consists of a series of voltages which represents samples of an analog wave that were taken at predefined times: A analog wave; B pulse amplitude modulation waveform

The level or amplitude of each sample is determined by a coder–decoder (codec) and a pulse proportional to the amplitude at the sampling instant is created. Figure 2.9A illustrates an analog wave that is to be digitized. The resulting signal, based upon the previously described sampling, is called a pulse amplitude modulation (PAM) wave which is illustrated in Figure 2.9B.

The PAM waves are a series of voltages that represent samples of the analog wave at defined times. Since they represent their information in an analog format, they are not suitable for transmission over relatively long distances. Due to this, the PAM waves must be quantized.

Quantization

The process of reducing a PAM signal to a limited number of discrete amplitudes is called quantization. This second part of the encoding process is necessary since the PAM samples can represent an infinite number of amplitude levels. Since the PAM waves represent a continuous height that can have any value within a range, while quantization results in a limited number of discrete amplitudes, this process will introduce errors. Such errors are called quantization noise, and represent the difference between the resulting discrete coded value of the PAM samples and the actual values of the samples. Thus, the number of discrete steps used to code PAM samples affects both the potential quantization error and the number of bits required to encode the resulting PCM data element.

Experiments have shown that the use of 2048 uniform quantizing steps provides the ability to obtain a sufficient capability to reproduce a voice signal of high quality. For 2048 quantizing steps, an 11-element code (2^{11}) would be required. Using a sampling rate of 8000 samples per second, the data rate required to digitize a voice channel would become 88 000 bps. This data rate is reduced by the third step in the PCM process—coding.

Coding

To reduce the number of quantum steps, two techniques are commonly used—non-uniform quantizing and companding prior to quantizing, followed by uniform quantizing. In non-uniform quantizing, the step assignments used for encoding are changed so that large steps are assigned to portions of high and low amplitudes, while smaller steps are assigned to intervals between those steps. In companding, the analog signal is first compressed prior to coding, followed by expansion after decoding. This permits a finer granularity in the form of additional steps to the smaller amplitude signals.

The objective of each technique is to reduce the number of quantum steps to 128 or 256, enabling either 7 bits ($2^7 = 128$ quantum steps) or 8 bits ($2^8 = 256$ quantum steps) to be used to encode each PAM sample.

Most PCM systems employ companding with non-uniform quantization. The compression, as well as eventual expansion, are based upon logarithmic functions that follow one of two laws: the A law and the 'mu' (μ) law.

Both North American and Japanese PCM systems employ μ law encoding, while European PCM systems utilize A-law encoding. Each technique defines the number of quantizing levels into which samples can be encoded and how those levels are arranged. The μ-law divides the quantization scale into 255 discrete units of two different sizes called segments or chords and steps. There are 16 segments or chords, eight used to represent positive signals and eight used to represent negative signals. Each chord is further subdivided into 16 steps. Since the zero level is shared, there are $16 \times 16 - 1$, or 255, levels available for use. In comparison, coding using the A law results in a 13-segment approximation since there are six segments and the segments passing through the origin are colinear and counted as one segment.

The curve for the A law is plotted from the formula

$$Y = \frac{AX}{(1+\log A)} \quad 0 \leq X \leq 1/A$$

The curve for the μ law is plotted from the formula

$$Y = \frac{\log(1 + \mu X)}{\log(1 + \mu)} \quad -1 \leq X \leq 1$$

The relationship between the linearity of an input signal and the resulting compressed input coded value of the signal is based upon the selection of values for the parameters A and μ. The selection of values for those parameters determines the range over which the ratio of signal-to-distortion remains relatively constant. This range is known as the dynamic range and has a value of approximately 40 dB. The dynamic range value is obtained using a value of 87.6 for A for the A law. For the μ law, μ is set to a value of 100 in older systems that use a seven-segment approximation of the logarithmic curve. For more modern 16-segment approximations, μ is set to a value of 255.

Figure 2.10 illustrates a general plot of quantization in the North American D2 PCM system. This system uses the μ law with a value of 255 assigned to μ. For clarity, only the positive portion of the curve is assigned values. Note that segment 6 has been exploded to illustrate that although the segments, which are also known as chords, are spaced logarithmically, within each chord the steps are spaced linearly.

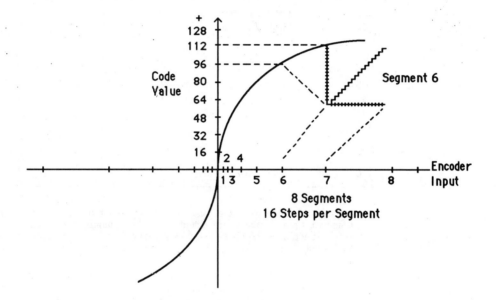

Figure 2.10 Quantizing curve used in North American D2 PCM systems. The North American D2 PCM system uses 16 segments (8 shown for clarity) spaced logarithmically based upon μ set to a value of 255

Table 2.1 lists the μ law PCM code table used in the North American D2 PCM system. In this system there are actually only 255 quantizing steps since steps zero and one use the same bit sequence to avoid a code sequence with no transitions, such as zeros only.

An examination of the entries in Table 2.1 illustrates the encoding format of PCM words. As indicated by the entries in the table, the first bit denotes whether or not the signal was above (1) or below (0) the horizontal axis. The next three bit elements identify the segment or chord, while the last four bits identify the

Table 2.1 North American D2 μ-law.

Code Level		Digit number							
		1	2	3	4	5	6	7	8
255	(Peak positive level)	1	0	0	0	0	0	0	0
239		1	0	0	1	0	0	0	0
223		1	0	1	0	0	0	0	0
207		1	0	1	1	0	0	0	0
191		1	1	0	0	0	0	0	0
175		1	1	0	1	0	0	0	0
159		1	1	1	0	0	0	0	0
143		1	1	1	1	0	0	0	0
127	(Center levels)	1	1	1	1	1	1	1	1
126	(Nominal zero)	0	1	1	1	1	1	1	1
111		0	1	1	1	0	0	0	0
95		0	1	1	0	0	0	0	0
79		0	1	0	1	0	0	0	0
63		0	1	0	0	0	0	0	0
47		0	0	1	1	0	0	0	0
31		0	0	1	0	0	0	0	0
15		0	0	0	1	0	0	0	0
2		0	0	0	0	0	0	1	1
1		0	0	0	0	0	0	1	0
0	(Peak negative level)	0	0	0	0	0	0	1*	0

*One digit is added to insure that the timing content of the transmitted pattern is maintained. Note that there are actually only 255 quantizing steps because steps '0' and '1' use the same bit sequence, thus avoiding a code sequence with no transitions (i.e., '0's only).

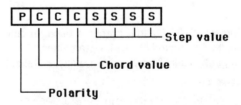

Figure 2.11 PCM word encoding format. The PCM word encoding format permits the polarity of the sample to be determined by the examination of one bit and prevents most line hits from significantly changing the value of the word

actual step or level within the chord. Figure 2.11 illustrates the PCM word encoding format which provides several advantages over a simple assignment based upon binary bit values. First, with a conventional encoding system, equipment would have to read eight bits to tell if a number was positive or negative. Using the encoding format illustrated in Figure 2.11 permits the polarity of the sample to be determined by the examination of one bit position. A second advantage of the PCM word encoding format illustrated in Figure 2.11 is that it prevents most line hits from significantly changing the value of the PCM word. As an example, the altering of a step bit would only affect the value of the step.

In comparison to the North American μ law, European systems quantize voice signals using a 13-segment approximation of the A law curve. For the most part, both techniques provide a similar level of quality for a reconstructed signal. In general, the A law produces a better signal-to-noise ratio at low levels, while the μ law has a lower idle channel noise level.

Companders

The term 'compander' is an acronym for the compressors and expanders used in communications carrier systems to improve the signal-to-noise ratio of an analog signal. Through the use of companders, the dynamic range of voice signals can be compressed prior to quantization. Then, after the encoded signal is converted back into its analog form by an interpolator, an expander expands the range of the analog signal to its original range. Figure 2.12 illustrates the companding process in block form.

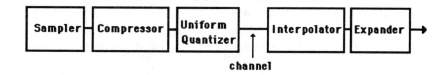

channel

Figure 2.12 The compressor–expander (companding) process. The compressor reduces the dynamic range of a voice signal prior to quantization, while the expander expands the range of the analog signal after the interpolator restores it to an analog format

Compressor unit operation   The compressor performs the following functions:

• It raises the power level of weak signals so they can be transmitted above the noise and crosstalk level associated with a typical communications channel.

• It attenuates very strong signals to minimize the possibility of crosstalk affecting other communications channels.

Figure 2.13 illustrates the typical operation of a compressor and expander. The compressor accepts a 60 db input power range, typically between 20 and 80 db. The compressor increases the weakest sounds in power from 20 to 40 db, while the strongest sounds are decreased in power from 80 to 70 db.

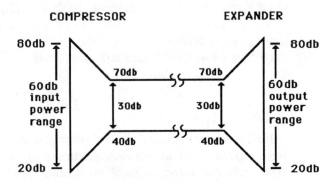

Figure 2.13 Compressor–expander operation. The compressor reduces the dynamic range of an analog signal by 30 dB, while the expander increases the signal to its original range

The expander unit works exactly opposite to the compressor unit, expanding the reduced power range back to its original form. Here the shift in power is opposite to that performed by the compressor, with the weakest sounds decreased in power from 40 to 20 db, while the strongest sounds increase in power from 70 db up to 80 db.

As a result of the compression quantization and encoding process, each PAM sample is encoded into either 7 data bits + 1 signaling bit used to indicate the polarity (sign) of the sample (128 quantizing steps) or 8 data bits (256 quantizing steps). For either encoding method, the resulting data rate becomes:

8000 samples/second × 8 bits/sample = 64 000 bps

Based upon the preceding, each voice conversation digitized according to the PCM method results in a data rate of 64 kbps. Since a North American T-carrier (called T1) supports 24 digitized voice conversations, its operating rate would appear to be 64 kbps × 24, or 1.536 Mbps.

In actuality, the T1 carrier in North America operates at 1.544 Mbps. The difference, which is 8000 bits per second, is used for framing to include synchronization. For European systems, the T-carrier supports 32 channels operating at 64 kbps, resulting in a data rate of 2.048 Mbps. Since signaling and synchronization are embedded in two 64 kbps channels, the actual usable data rate of European systems is 64 kbps × 30, or 1.92 Mbps. Both North American T1 carrier framing and European signaling and synchronization are covered in detail in Chapter 4.

2.2.2 Adaptive differential pulse code modulation (ADPCM)

Adaptive differential pulse code modulation (ADPCM) is the only technique recommended by the CCITT as a worldwide standard, G.721. ADPCM uses a sampling rate of 8000 samples per second, the same as used in PCM. A transcoder uses an algorithm to reduce the number of quantizing levels to 16, permitting each sample to be represented by a 4-bit word. This process, which is illustrated in the left portion of Figure 2.14, results in the predicted value being subtracted from the actual value, permitting the difference to be transmitted in the form of a 4-bit PCM word.

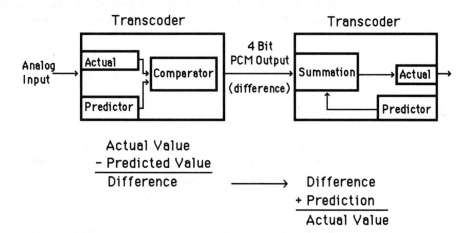

Figure 2.14 Adaptive differential pulse code modulation operation. In adaptive differential pulse code modulation (ADPCM), the predicted value is subtracted from the actual PCM value resulting in the difference being transmitted as a 4-bit word

The 4-bit word used by ADPCM represents the difference between the actual value of the signal and its predicted value. This difference information is sufficient to reconstruct the amplitude of the signal.

At the opposite end of the channel, another transcoder with an identical predictor performs the ADPCM process in reverse, restoring the predicted signal to the original 8-bit code. As indicated in the right portion of Figure 2.14, a summation process adds the received difference to the predicted value generated by the receiving transcoder to obtain the actual value of the signal.

The key to ADPCM is the fact that the human voice does not change significantly from one sampling interval to another, permitting a high degree of prediction accuracy. This means that the difference between the predicted and actual signal is very small and can be encoded using only 4 bits. In the event that successive samples should vary widely, the algorithm used for prediction will adapt to the changes by increasing the range represented by the 4 bits. However, as might be expected, this adaptation reduces the accuracy of voice frequency reproduction.

ADPCM results in the data rate of a digitized voice conversation becoming:

$$8000 \text{ samples/second} \times 4 \text{ bits/sample} = 32\,000 \text{ bps}$$

Note that ADPCM results in a data rate one-half of the PCM data rate. Thus, the use of ADPCM doubles the digitized voice carrying capacity of a North American T1 system to 48 channels and a European T1 system (called E1) to 60 voice channels.

Advantages

The major advantage of ADPCM is that its use permits the capacity of voice and voice/data systems to increase. This is because digitized voice can be reduced to 32 kbps, permitting a doubling of the number of voice conversations that can be carried on a system or lowering the data rate to support voice which allows data transmission support to increase. In addition, ADPCM requires less complex circuitry than other 8- to 4-bit encoding schemes, such as time assigned speech interpolation (TASI), continuous variable slope delta modulation (CVSD), and near instantaneous companding (NIC).

Disadvantages

Like most technology, the use of ADPCM involves several trade-offs. Until recently, transcoding algorithms were not standardized. Thus, the system of one manufacturer may not be compatible with those manufactured by another vendor. Even if the systems are compatible, it may not be a good idea to mix and match devices using the same ADPCM compression algorithms since using equipment from different vendors inhibits the uniformity of diagnostic capabilities. In addition, because the patterns of voice and data differ, separate adaptive predictors must be dedicated to each, requiring a speech/data detector in each transcoder to determine which algorithm to employ.

The last major disadvantage associated with the use of ADPCM relates to its ability to transmit modulated data. In general, ADPCM provides marginal support for 4-wire modems operating above 4800 bps since signal-to-noise level resulting from the application of a predictive algorithm is not constant.

2.2.3 Continuous variable slope delta modulation (CVSD)

Continuous variable slope delta modulation (CVSD) was originally used in military communications as it facilitated the encryption of analog voice conversations. Today, several T-carrier multiplexer vendors market CVSD digitization modules that can be added to their equipment to digitize voice at relatively low data rates.

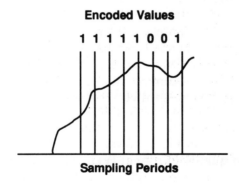

Figure 2.15   Continuous variable slope delta modulation (CVSD). Samples that have an increased height in comparison to a previous sample are encoded as a '1', while a sample that has a height less than a previous sample is encoded as a '0'

In the CVSD digitization technique, the analog input voltage is compared to a reference voltage. If the input is greater than the reference, a '1' is encoded, while a '0' is encoded if the input voltage is less than the reference level. This permits a 1-bit data word to represent the digitized voice signal. Figure 2.15 illustrates the resulting encoded values of a portion of an analog signal at defined sampling periods.

Early military systems sampled the analog waveform 8000 times per second, resulting in a bit rate of 8 kbps. Although some commercial systems can also be set to that sampling rate, in general, good quality voice reproduction requires a faster sampling rate. Today, most military and commercially available CVSD systems sample the analog input at 16 000, 24 000, or 32 000 times per second, resulting in a bit rate of 16 kbps, 24 kbps, or 32 kbps to represent a digitized voice signal.

Table 2.2   T-carrier voice capacity using CVSD.

CVSD data rate (kbps)	North America	Europe
16	96	120
24	72	90
32	48	60

Table 2.2 lists the maximum number of voice channels North American and European T-carrier systems can support based upon commonly used CVSD digitization rates. As might be expected, the large voice channel support at low CVSD digitization rates is the major advantage obtained from this digitization method. Unfortunately, the use of CVSD modules in multiplexers is not recommended for passing modem modulated data in a digital format. Due to the problem associated with carrying modulated data, communications carriers do not use this digitization technique in their facilities.

2.2.4 Variable quantizing level (VQL)

Variable quantizing level (VQL) is a voice digitization technique developed by Aydin Monitor Systems to provide a 32 kbps digital stream from a telephone voice frequency (VF) channel. One of the most interesting aspects of this digitization technique is its use of 4 error control bits which enables the receiving end to detect and correct any single bit error that might occur in the header of a 'VQL word'. This limited measure of protection provides partial protection to the digitization's critical component, which is the header of the VQL word.

Under the VQL algorithm, the speech waveform is filtered from 3400 to 3000 Hz. To reduce the high end of the voice passband, the resulting passband is sampled 6667 times per second, and PCM encoded samples are then processed in blocks of 40, which corresponds to a 6 millisecond snapshot of the speech signal. Since there are 6667 samples per second within the 3000 Hz bandwidth, the ratio of sampling to bandwidth is approximately 2.2. This ratio exceeds PCM's 2.0 ratio (8000/4000), which provides better fidelity of voice reproduction within the 300 to 3300 Hz frequency range.

For each block of 40 PCM encoded samples, a maximum amplitude is obtained which is then divided into 11 steps of equal magnitude. Each sample in the block is then compared to the maximum amplitude and assigned a new code which corresponds to the nearest of the 11 levels as illustrated in Figure 2.16. The data for a 6 millisecond block is then formed into a 'VQL word' which contains the maximum amplitude of the 40 samples, signaling information, forward error correction bits, and the 40 encoded samples. Due to the blocking of samples into a VQL word, an inherent delay up to 6 milliseconds is associated with this digitization technique.

In the example illustrated in Figure 2.16, assume block I in a sequence of 40 PCM samples had the maximum amplitude of all samples. Then, the height of that block, which is 44, is converted into a sequence of 11 levels. Each level is then encoded by the use of 4 bits plus a sign bit, resulting in each sample being represented by 5 bits. Next, the maximum amplitude of the sequence of 40 samples in the block, which in this instance is 44, is encoded into a header, followed by 2 signaling bits and 4 forward error correcting bits.

The 4 forward error correcting bits protect the maximum amplitude and 2 signaling bits, permitting the receiving end to detect and correct any single bit error in the 12-bit header.

Based upon the preceding, the VQL channel rate would be computed as follows:

$$\frac{6667 \text{ samples/second}}{40 \text{ samples/block}} = \frac{166\,675 \text{ blocks}}{\text{second}}$$

40 samples × 5 bits/sample + 12 header bits = 212 bits block

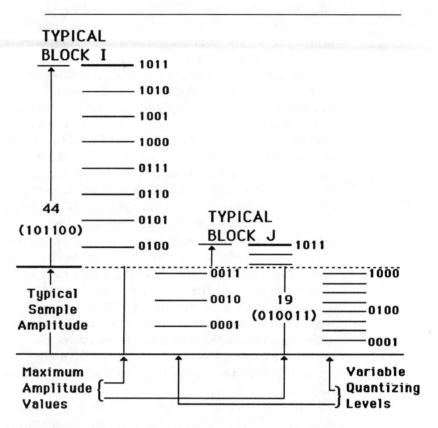

Figure 2.16 VQL digitization. Forty samples are converted to an 11-step scale defined by the maximum amplitude in the block of samples

$$166\,675 \text{ blocks/second} \times 212 \text{ bits/block} = 35\,335 \text{ bps}$$

Since this data rate exceeds 50 percent of the conventional PCM data rate, additional processing is used to obtain a 32 kbps rate that doubles the capacity of a T-carrier. To obtain a 32 kbps channel rate, two samples are combined to improve coding efficiency. Since each VQL sample can assume 22 possible values (11 positive and 11 negative steps), two samples taken together can assume one of 22 × 22, or 484, possible values. That product can be encoded in 9 bits ($2^9 = 512$), which is equivalent to 4.5 bits per sample.

Using 4.5 bits per sample shortens the VQL word to

40 samples $\times$ 4.5 bits/sample + 12-bit header = 192bits

Then

166 67 blocks/second $\times$ 192 bits/VQL word = 32 kbps

which is exactly one-half of the conventional PCM data rate.

Advantages and disadvantages

Due to the encoding techniques used by VQL, as the signal amplitude decreases, the magnitude of the quantizing steps decreases almost in proportion. This results in a near-constant signal-to-noise level, which permits support of four-wire modems operating up to 9600 bps. In addition, the use of forward error correcting provides a measure of integrity to the most critical part of the encoding process, the VQL word header. Thus, VQL has a limited ability to correct itself from line hits and other disturbances.

Although VQL's performance from 300 to 3000 Hz is excellent, its design results in the loss of the high frequencies between 3000 and 3400 Hz. While the loss of high frequencies does not appreciably affect voice, it can significantly reduce the operating rate of packetized ensemble protocol modems that attempt to transmit on up to 512 carriers spaced throughout the voice channel passband. Thus, this voice encoding technique can adversely affect some modem modulation methods used to transmit data.

Another problem with VQL, as with many other voice digitization techniques, is that this method of voice encoding is not standardized.

2.2.5 Digital speech interpolation (DSI)

Digital speech interpolation (DSI) can be considered as a form of statistical voice multiplexing, since it takes advantage of the idle moments spread throughout a normal telephone conversation. In this voice digitization method, only active channels are digitized and transmitted, permitting the equipment to take advantage of the half-duplex nature of voice conversations, as well as the natural pauses inherent in speech.

DSI efficiency is based upon the fact that, for a large group of active channels, the long-term idle time per channel is between

60 and 65 percent of the call duration. The interpolation process operates by filling the gaps in some channels with speech content simultaneously present in other channels.

The ratio of user channels to voice transmission channels is known as the DSI gain. The DSI gain can be as high as 2.5 to 1 for a very large group of user channels. However, for T-carrier circuits that have a relatively small number (24 or 30) of voice channels, a DSI gain of 2 to 2.2 is used in multiplexer voice modules. Since a speaker's activity is not predictable, the DSI technique becomes more efficient as the number of conversations increases.

Figure 2.17 illustrates an example of the DSI encoding process. Each active channel, denoted by the greater than (>) sign, is digitized and placed into a transmission block. The first field in the block, which is labeled interpolation control, denotes the active channel numbers or addresses of the digitized voice samples that follow in the block. In this example, the interpolation control field would contain channel addresses 3, 5, ..., 96.

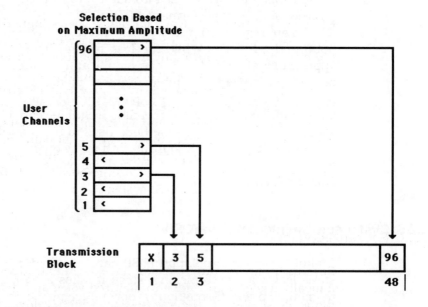

X = Interpolation Control

Figure 2.17 DSI encoding process. The interpolation control header contains the address of the active channels that are encoded and contained in the transmission block

Most DSI techniques use either ADPCM or VQL digitization, resulting in the effective data rate per voice channel being reduced to 16 kbps. This data rate assumes a DSI gain of two. Thus,

$$\frac{32\,000 \text{ bps}}{2 \text{ DSI gain}} = 16\,000 \text{ bps}$$

Based upon the preceding, the use of DSI with ADPCM or VQL digitization permits up to 96 voice channels to be carried on a North American T-carrier or 120 channels on a European T-carrier facility.

The primary advantage of DSI is its ability to provide a 4 : 1 ratio over the capacity of a T-carrier carrying PCM encoded voice when DSI is used in conjunction with ADPCM or VQL. To understand the major disadvantage of DSI, assume that in a 6-ms time period there are more than 44 user channels with speech energy that meet the employed selection criteria. This will result in some channels not being transmitted, causing a random 6-ms discontinuity or dropout. This dropout, which results in a cutoff of speech energy, is more commonly known as clipping.

Dropouts are more likely to occur in weak signals since the DSI selection algorithm normally favors higher-level signals. Sometimes, dropouts of 18 or 24 ms can be inaudible; however, the human ear can normally detect dropouts in excess of 36 ms.

To prevent clipping or extended dropouts, DSI employs an algorithm to distribute small dropouts evenly over the population of active channels. Unfortunately, this would cause havoc to digital data carried by a DSI system. Thus, to protect data, DSI devices assign a channel carrying modem or digital data traffic to a fixed slot. This shields the data from the interpolation process and insures that dropouts will not occur; however, it also decreases the overall efficiency of a DSI system.

2.2.6 Packetized voice

Several T-carrier multiplexers have been introduced that dynamically assign bandwidth to digitized voice and data. This technique is similar to the method by which value-added carriers packetize data; however, unlike conventional packet networks, T-carrier multiplexers employing this method take advantage of the silent periods of both voice conversations and data transmission.

When a packetized voice system is used, the T-carrier channel is considered to be a virtual circuit, resulting in both voice and data packets flowing between multiplexers. By using ADPCM to compress voice into a 32 kbps digital data stream prior to packetization, up to 96 voice channels can be supported on a 1.544 Mbps T-carrier link or 120 voice channels on a 2.048 Mbps T-carrier facility.

One of the major problems associated with packetized voice is the dynamic routing of packets through a series of multiplexers in a network. Due to the time required to examine packet headers and route packets through a network, packets may arrive at staggered intervals of 20 to 25 milliseconds or more from one another, causing a hearing annoyance to the distant listener. The reader is referred to Chapter 6 in which different types of multiplexer operations are discussed for additional information on the operation and utilization of multiplexers that packetize voice and data.

2.2.7 Linear predictive coding (LPC)

The basis behind the operation of linear predictive coding is the analysis it performs on analog speech. Figure 2.18 illustrates the speech producing elements of the human vocal track. During LCP encoding, the analog voice input is analyzed and then converted into a set of digital parameters for transmission. At the receiver, a synthesizer recreates an analog voice output based upon the received set of digital parameters. By limiting the analysis of the voice signal to four sets of voice parameters, a very low data rate can be used to transmit voice data in digital form.

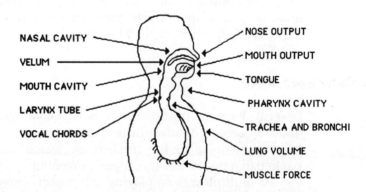

Figure 2.18 Speech producing elements of the human vocal tract. The speech producing elements of the human vocal tract are analyzed by a linear predictive encoder to synthesize a voice conversation

Operation

In linear predictive coding, the voice signal is first sampled by a 12-bit analog-to-digital converter. The output of the converter is then used as input to four parametric detectors. A pitch detector analyzes the data to obtain the fundamental pitch frequency at which vocal cords vibrate. Next, a voice/unvoiced detector senses whether sound is caused by the vibration of vocal cords (voice) or by sounds such as 'shhh' (unvoiced) that do not vibrate. A power detector then determines the amplitude (volume or loudness) of the sound, while a spectral data decoder models the resonant cavity formed by the throat and the mouth.

Since LPC sends speech parameters rather than the amplitude of waveforms, it is actually a method of speech synthesis.

Equipment application

LPC has been integrated into several devices collectively known as voice digitizers that can operate at 2400 or 4800 bps. Typically, data from the four LPC detectors is stored in a 54-bit buffer that is released every 22.5 milliseconds. This results in 44.444 samples per second which, when multiplexed by 54 bits per sample, permits an analog voice conversation to be digitized at 2400 bps.

Figure 2.19 illustrates two typical voice digitizer applications. In Figure 2.19A, two voice digitizers are connected to a PBX, enabling the output of both devices to be multiplexed onto a common circuit. Using this technique, organizations can derive two or more voice channels from one analog circuit.

In Figure 2.19B, the integration of one voice channel into a data network is illustrated. In this example, the output of the voice digitizer is multiplexed with digital data sources onto a common data channel.

Constraints to consider

The primary constraints associated with the use of devices that employ linear predictive encoding are the cost of equipment and the fidelity of the reconstructed signal. Currently, the cost of voice digitizers exceeds $7000 per unit, making their use economically feasible only for long distances. Due to this, the primary use of voice digitizers is to derive multiple voice conversations on expensive international circuits.

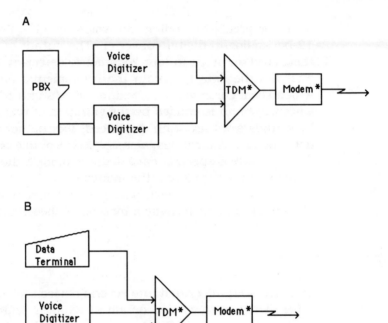

Figure 2.19 Voice digitizer applications: A multiple voice channel derivation; B mixed voice and data networks; *the time division multiplexer (TDM) and modem can be replaced by a multipoint modem that contains a built-in synchronous time division multiplexer

With respect to the fidelity of reconstructed voice, LPC synthesizes both the speaker's conversation and any background noise. Due to the method used to synthesize voice conversations, background noise is accentuated, which can result in some disturbance to the reconstructed analog signal. Thus, the use of LPC is more suitable to conversations originating in an office environment than for conversations on a factory floor where there may be a large amount of background noise.

A second fidelity problem associated with LPC is its low bit rate. This results in a small bit error rate affecting a larger portion of speech quality than an equivalent error rate on a PCM or ADPCM system. In some situations, a low bit error rate can actually result in the reconstructed voice appearing as speaker stuttering.

2.3 TELEPHONE COMPANY TRANSMISSION HIERARCHY

In this section we will examine the evolution of transmission system facilities used by communications carriers to interconnect their central offices. In doing so we will first focus our attention upon the use of frequency division multiplexing (FDM), which was the earliest technique used to enable multiple simultaneous voice conversations to be routed onto a common circuit interconnecting two carrier central offices. This examination of the use of FDM equipment will be followed by a short review of the operation of time division multiplexing (TDM) equipment whose utilization contributed directly to the development of T-carrier facilities. Based upon the preceding, we will conclude this section by examining the evolution of T-carrier systems and their use by communications carriers to interconnect their central offices.

To reduce the number of physical lines required to connect telephone company offices to one another, communications carriers employ multiplexing. Originally, frequency division multiplexing was used exclusively by communications carriers. Gradually, FDM has been replaced by time division multiplexing that utilizes the T-carrier as a digital transport mechanism. This evolution to TDM equipment and T-carrier facilities was based upon the advantages associated with digital signaling in comparison to analog signaling as previously covered in this chapter.

2.3.1 Frequency division multiplexing

Employing frequency division multiplexing between carrier central offices requires the use of a communications circuit that has a relatively wide bandwidth. This bandwidth is then divided into subchannels by frequency. When a communications carrier uses FDM for the multiplexing of voice conversations onto a common circuit, the 3 kHz passband of each conversation is shifted upward in frequency by a fixed amount of frequency. This frequency shifting places the voice conversation into a predefined channel of the FDM multiplexed circuit. At the opposite end of the circuit, another FDM demultiplexes the voice conversations by shifting the frequency spectrum of each conversation downward in frequency by the same amount of frequency in which it was previously shifted upward.

As previously mentioned, the primary use of FDM equipment by communications carriers was to enable those carriers to carry a large number of simultaneous voice conversations on a common circuit routed between two carrier offices. The actual process for allocating the bands of frequencies to each voice conversation has been standardized by the CCITT. CCITT FDM recommendations govern the channel assignments of voice multiplexed conversations based upon the use of 12, 60, and 300 derived voice channels.

CCITT FDM recommendations

The standard group as defined by CCITT recommendation G.232 occupies the frequency band from 60 to 108 kHz. This group can be considered as the first level of frequency division multiplexing and contains 12 voice channels, with each channel occupying the 300 to 3400 Hz spectrum shifted in frequency.

The standard supergroup as defined by CCITT recommendation G.241 contains five standard groups, equivalent to 60 voice channels. The standard supergroup can be considered as the second level of frequency division multiplexing and occupies the frequency band from 312 to 552 kHz.

The third CCITT FDM recommendation, known as the standard mastergroup, can be considered as the top of the FDM hierarchy. The standard mastergroup contains five supergroups. Since each supergroup consists of 60 voice channels, the mastergroup contains a total of 300 voice channels. The standard mastergroup occupies the frequency band from 812 to 2044 kHz. Figure 2.20 illustrates the three standard CCITT FDM groups, as well as the relationship between groups.

2.3.2 Time division multiplexing

The fundamental operating characteristics of a TDM are shown in Figure 2.21. Here, each low to medium speed digital data source is connected to the multiplexer through an input/output (I/O) channel adapter. The I/O adapter provides the buffering and control functions necessary to interface the low to medium speed data sources to the multiplexer. Within each adapter, a buffer or memory area exists which is used to compensate for the speed differential between the data sources and the multiplexer's internal operating speed. Data is shifted from the terminal to the I/O adapter at different rates, depending upon the speed of the

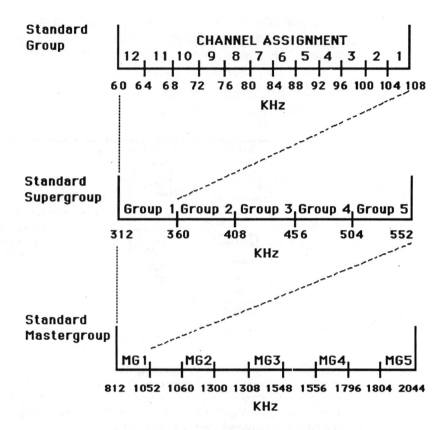

Figure 2.20 Standard CCITT FDM groups. CCITT FDM recommendations govern the assignment of 12, 60, and 300 voice channels on wideband analog circuits

connected input data sources; but when data is shifted from the I/O adapter to the central logic of the multiplexer, or from central logic to the composite adapter, it is at the much higher fixed rate of the TDM. On output from the multiplexer to each data source the reverse is true, since data is first transferred at a fixed rate from central logic to each adapter and then from the adapter to the attached device at the data rate acceptable to the device.

The central logic of the TDM contains controlling, monitoring, and timing circuitry which facilitates the passage of individual terminal data to and from the high-speed transmission medium. The central logic will generate a synchronizing pattern which is used by a scanner circuit to interrogate each of the channel adapter buffer areas in a predetermined sequence, blocking the bits of characters from each buffer into a continuous, synchronous data stream which is then passed to a composite adapter. The composite adapter contains a buffer and functions similar to the I/O channel adapter. However, it now compensates for the difference in speed between the high-speed transmission medium and the internal speed of the multiplexer.

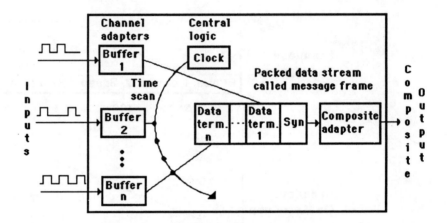

Figure 2.21 Time division multiplexing. In time division multiplexing, data is first entered into each channel adapter buffer area at a transfer rate equal to the device to which the adapter is connected. Next, data from the various buffers are transferred to the multiplexer's central logic at the higher rate of the device for packing into a message frame for transmission

TDM techniques

The two TDM techniques available are bit interleaving and character interleaving. Bit interleaving is generally used in systems which service synchronous devices, whereas character interleaving is generally used to service asynchronous devices. When interleaving is accomplished on a bit-by-bit basis, the multiplexer takes one bit from each channel adapter and then combines them as a word or frame for transmission. As shown in Figure 2.22A, this technique produces a frame containing one data element from each channel adapter.

When interleaving is accomplished on a character-by-character basis, the multiplexer assembles a full character from each data source into a frame for transmission as shown in Figure 2.22B.

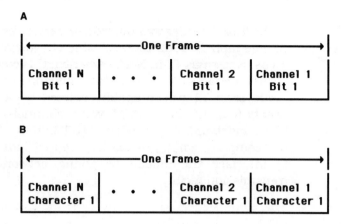

Figure 2.22 TDM techniques. In bit interleaving, a frame is assembled by the TDM gathering one bit from each input data source. In character interleaving, the multiplexer assembles the frame by gathering one character from each input data source: A bit interleaving; B character interleaving

In general, bit interleaving is more efficient than character interleaving as it minimizes the potential delay in the data flow of multiplexed synchronous events. Unfortunately, bit interleaving cannot be used with some types of digital networking activities, such as digital access cross connect (DACC) systems, that operate on an 8 bit per character basis. To compensate for this dilemma, some multiplexer vendors have introduced equipment that operates on a combined bit and character interleaving basis. That is, a portion of the composite bandwidth, or more accurately a portion of the multiplexing frame, is reserved for character interleaving while the remainder of the multiplexer frame is reserved for bit interleaving.

2.3.3 T-carrier evolution

T-carrier facilities were originally developed by telephone companies as a mechanism to relieve heavy loading on interexchange circuits. First employed in the 1960s for intra-carrier communications, T-carrier facilities only became available to the general public within the last 10 years as a commercial offering.

Since the mid-1980s, reductions in T-carrier tariffs have made them into an economic data plus voice transportation highway. Currently, T-carrier multiplexers are exhibiting a sales growth of 25 to 30 percent per year, among the highest rate of growth of all categories of communications equipment.

The first T-carrier was placed into service by American Telephone & Telegraph in 1962 to ease cable congestion problems in urban areas. Known as T1 in North America, this wideband digital carrier facility operates at a 1.544 Mbps signaling rate.

The term T1 was originally defined by AT&T and referred to twenty-four 64 kbps PCM voice channels carried in a 1.544 Mbps wideband signal. When AT&T initiated use of its T1 carrier, the company employed digital channel banks which were used to interface the analog telephone network to the T1 digital transmission facility.

Channel banks

Channel banks used by telephone companies were originally analog devices. They were designed to provide the first step required in the handling of telephone calls that originated in one central office, but whose termination point was a different central office. The analog channel bank included frequency division multiplexing equipment, permitting it to multiplex, by frequency, a group of voice channels routed to a common intermediate or final destination over a common circuit. This method of multiplexing was previously illustrated in Figure 2.20.

The development of pulse code modulation resulted in analog channel banks becoming unsuitable for use with digitized voice. AT&T then developed the D-type channel bank which actually performs several functions in addition to the time division multiplexing of digital data.

The first digital channel bank, known as D1, contained three key elements as illustrated in Figure 2.23. The codec, an abbreviation for coder–decoder, converted analog voice into a 64 kbps PCM encoded digital data stream. The TDM multiplexes 24 PCM encoded voice channels and inserts framing information to permit the TDM in a distant channel bank to be able to synchronize itself to the resulting multiplexed data stream that is transmitted on the T1 span line. The line driver conditions the transmitted bit stream to the electrical characteristics of the T1 span line, insuring that the pulse width, pulse height and pulse voltages are correct. In addition, the line driver converts the unipolar digital signal transmitted by the multiplexer into a bipolar signal suitable for transmission on the T1 span line. Due to the operation of the digital channel bank, this equipment can be viewed as a bridge from the analog world to the digital world. To insure the quality of the resulting multiplexed digital signal, AT&T installed repeaters at intervals of 6000 feet on span lines constructed between central offices. Although repeaters are still required on local loops to a subscriber's premises and on copper wire span lines, the introduction of digital radio and fiber optic transmission has added significant flexibility to the construction and routing of T-carrier facilities.

Improvements made to the encoding method used by the D1 channel bank resulted in the development and installation of the D1D and D2 channel banks. Channel banks currently used include D3, D4, and D5, whose simultaneous voice channel carrying capacity varies from 24 to 96 channels. Additional information concerning the operation of channel banks is contained in Chapter 6.

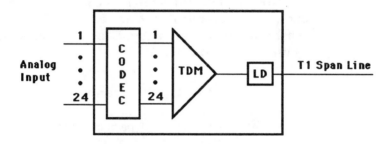

Figure 2.23 The D1 channel bank: TDM = time division multiplexer, LD = line driver

Today, T1 lines are available from a variety of communications carriers, including AT&T, MCI, US Sprint, and others. In Europe, the equivalent T1 carrier, which is known as E1 and CEPT PCM-30, is available in most countries under different names. As an example, in the United Kingdom E1 service is marketed under the name MegaStream.

Framing structure overview

In North America, the T1 carrier was designed to support the transmission of 24 channels of digitized voice.

Each channel is sampled 8000 times per second and 8 bits are used to represent the encoded height of the sampled analog wave. For synchronization, as well as other functions that are discussed in Chapter 3, one framing bit is added to the digitized multiplexed data that represents 24 PCM encoded voice conversations.

Thus,

$$8 \text{ bits} \times 24 \text{ channels} + 1 \text{ framing bit} = 193 \text{ bits/frame}$$

Since 8000 frames are transmitted each second, the bit rate is

$$193 \text{ bits/frame} \times 8000 \text{ frames/second} = 1.544 \text{ Mbps}$$

which is the operating rate of the North American T1 carrier facility.

In Europe, the T-carrier is commonly referred to as an E1 facility or a CEPT PCM-30 system, where CEPT is an acronym for the Conference of European Postal & Telecommunications, a European standards organization.

CEPT uses a 32-channel system, where 30 channels are used to transmit digitized speech received from incoming telephone lines, while the remaining two channels are used for signal and synchronization information. Each channel is assigned a time slot as listed in Table 2.3.

Table 2.3 CEPT time slot assignments.

Time slot	Type of information
0	Synchronization (framing)
1–15	Speech
16	Signaling
17–31	Speech

The frame composition of an E1 or CEPT system consists of 32 channels of 8 bits per channel, or 256 bits per frame. No framing information is required to be added to the frame as in the North American T1 carrier since synchronization is carried separately in time slot zero.

Since 8000 frames per second are transmitted, the bit rate of an E1 facility becomes

$$256 \text{ bits/frame} \times 8000 \text{ frames/second} = 2.048 \text{ Mbps}$$

Similar to the hierarchy illustrated for FDM in Figure 2.20, communications carriers have developed a hierarchy of digital carrier levels. Table 2.4 lists the digital carrier hierarchy levels in North America, Europe, and Japan.

Table 2.4 Digital hierarchy levels.

North America

Line type	Line signal standard	Number of voice circuits	Bit rate (Mbps)
T1	DS1	24	1.544
T1C	DS1C	48	3.152
T2	DS2	96	6.312
T3	DS3	672	44.736
T4	DS4	4032	274.176
T5	DS5		
T6	DS6		

Europe

Level number	System	Number of voice circuits	Bit rate (Mbps)
1	M1	30	2.048
2	M2	120	8.448
3	M3	480	34.368
4	M4	1920	139.264
5	M5	7680	565.148

Japan

Level number	System	Number of voice circuits	Bit rate (Mbps)
1	F-1	24	1.544
2	F-6M	96	6.312
3	F-32M	480	34.064
4	F-100M	1440	97.728
5	F-400M	5760	397.20
6	F-4.6G	23040	1588.80

In North America, T1C and T2 were developed to boost the carrying capacity of copper wire pairs beyond that obtained by using T1. These facilities are primarily restricted to use by telephone companies. T3, which operates at 44.736 Mbps, initially was offered to commercial organizations during 1988 and can be expected to grow in use.

The line types listed in the top portion of Table 2.4 actually reference the type of signal each line is capable of carrying. A T1 line carries a DS1 signal. Here DS1 (digital signal, level 1) is the 1.544 Mbps signal defined by AT&T to include pulse height and width,

54 _____ FUNDAMENTALS OF DIGITAL NETWORKING

impedance, and other parameters. Although column two labeled 'Line signal standard' commences with DS1, in actuality, the lowest signal level in the digital hierarchy is DS0 (digital signal, level 0). DS0 refers to each 64 kbps digitized PCM data stream generated in a D-type channel bank, with 24 such DS0 channels along with framing bits used to form a DS1 signal.

Figure 2.24 illustrates the North American digital signal hierarchy. Note that DS0 originates at the digital channel bank located in the lower left portion of the illustration. The five data rates shown entering the data multiplexer in the upper left portion of Figure 2.24 are Dataphone Digital Services transmission facilities, which are also commonly referred to as subrate services as they operate below the DS1 rate. Dataphone Digital Services was originally developed by AT&T as an all-digital transmission facility for the transmission of data as opposed to voice. Since its introduction by AT&T, other communications carriers have introduced equivalent digital facilities for the transmission of data. The characteristics and operation of AT&T's DDS transmission facility, as well as other digital services, are discussed in Chapter 3.

In the North American digital signal hierarchy, the M12 multiplexing system used by AT&T accepts four DS1 input signals and produces a DS2 output representing 96 DS0 channels operating at 6.312 Mbps. The AT&T M13 multiplexer operates upon 28 DS1 inputs, while that carrier's M23 multiplexer operates upon seven DS2 inputs, with both devices generating a DS3 output operating at 44.736 Mbps which represents 672 DS0 channels. The highest order AT&T multiplexer, which is the M34 device, accepts six DS3 inputs to form one DS4 output operating at 274.176 Mbps which represents 4032 DS0 channels.

REVIEW QUESTIONS

1 What is the difference between unipolar non-return to zero signaling and unipolar return to zero signaling?

2 Why is unipolar return to zero signaling easier to sample than unipolar non-return to zero signaling?

3 Why must line sampling be used to determine the value of each bit when polar non-return to zero signaling is used?

4 What are two advantages of bipolar return to zero signaling?

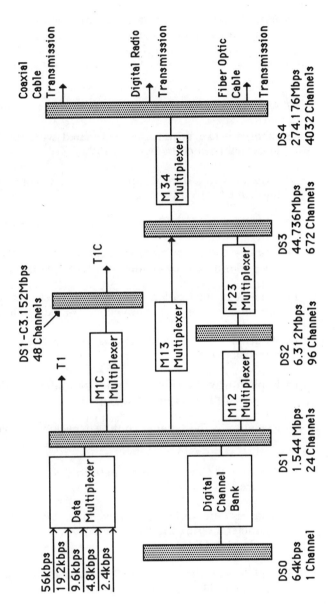

Figure 2.24 North American digital signal hierarchy

5 Why is a bipolar alternate mark inversion signaling technique used on digital transmission facilities?

6 Discuss the function of each of the three steps in the pulse code modulation process.

7 What is quantization noise and how does it occur?

8 What are two advantages associated with the PCM word encoding format illustrated in Figure 2.11?

9 Explain why ADPCM operates at 32 kbps and the advantages associated with this data rate.

10 Discuss several disadvantages associated with the use of ADPCM.

11 What is the major problem associated with CVSD modulation that precludes its use by communications carriers?

12 What voice digitization technique provides a measure of protection to the encoded signal that protects it from a small impairment?

13 What is clipping and how is it prevented under digital speech interpolation (DSI)?

14 Explain why the efficiency of a DSI system significantly decreases as its data traffic carrying use increases.

15 Explain why the dynamic routing of packetized voice through a complex network can result in a hearing annoyance to the distant location.

16 What are the two primary constraints associated with the use of voice digitizers?

17 What is the difference between a bit interleaving TDM and a character interleaving TDM? When would you prefer to use a character interleaving TDM?

18 Explain why the E1 or CEPT 30 system does not require a separate framing bit like the North American T1 carrier.

3

DIGITAL
NETWORKING
FACILITIES

In this chapter we will examine the operation and utilization of several types of North American and European digital network facilities. This examination will first focus upon subrate facilities which can be defined as a digital network facility operating at or below the DS0 data rate of 64 kbps. Since a grouping of 24 or 32 DS0 channels was originally multiplexed onto a North American or European T-carrier facility, our next focus of attention will be upon T-carrier transmission facilities. This will be followed by an overview of the method used by communications carriers to switch DS0 channels between T-carriers, a process known as digital access cross connect (DACS). This overview of DACS will include how it can be used by end-user organizations to control the routing of voice and data within an end-user's network. Concluding this chapter will be an examination of the most recent type of digital facility to be commercially marketed—fractional T1 or FT1.

3.1 SUBRATE FACILITIES

3.1.1 Dataphone Digital Service (DDS)

A subrate facility can be classified as a digital network communications line that operates at a data rate less than a T-carrier data rate—1.544 Mbps in North America and 2.048 Mbps in Europe. Two commercial examples of digital subrate facilities are AT&T's Dataphone Digital Service and British Telecom's KiloStream service offering.

Dataphone Digital Service (DDS) was approved by the US Federal Communications Commission in December, 1974. Currently, there are over 100 cities in the United States that are connected to the DDS network as well as connections which provide an interconnection capability to digital networks operated by other US and foreign communications carriers.

DDS is an all-synchronous facility. Currently supported transmission rates are listed in Table 3.1. For transmission at different data rates, specialized equipment to include multiplexers and/or converters must be employed.

Table 3.1   DDS offerings.

Data rate (kbps)	Type of service
2.4	leased
4.8	leased
9.6	leased
19.2	leased
56.0	leased
56.0	switched

Carrier structure

DDS facilities are routed from a subscriber's location to an office channel unit (OCU) located in the carrier's serving central office. Since there is a variety of multiplexing methods that can be employed by a communications carrier to combine DDS facilities onto a T-carrier, let us focus our attention upon two methods that will illustrate the relationship between DDS and a T1 circuit. Figure 3.1 illustrates the multiplexing arrangement within an AT&T serving central office that supports DDS transmission at 9.6 kbps and 56 kbps.

The data service units (DSUs) at the subscriber's location can be viewed as 'digital modems' since they modulate the unipolar signal received from data terminal equipment, including computer ports, multiplexer ports and terminal ports, into a modified bipolar signal suitable for transmission on the DDS network. Originally, separate channel service units (CSUs) and data service units were required

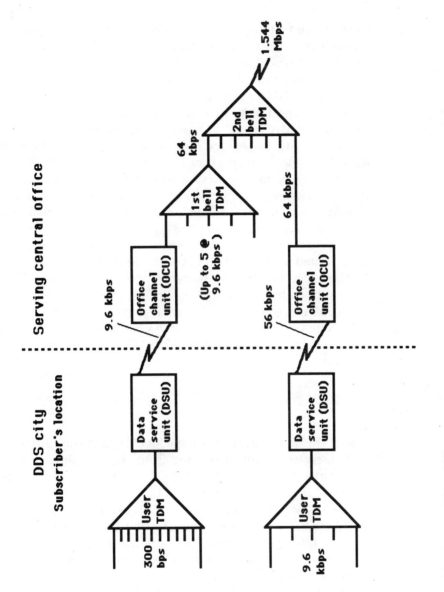

Figure 3.1   DDS multiplexing arrangement

to interface equipment to the DDS network. Today, many vendors manufacture DSUs that, in effect, combine the functions performed by separate DSUs and CSUs. The operation of both CSUs and DSUs is covered later in this chapter.

The user TDMs shown in Figure 3.1 illustrate two methods by which end-users can transmit data to maximize the data handling capacity of different DDS facilities. The user TDM shown in the upper left corner of Figure 3.1 illustrates how asynchronous transmission can be supported on DDS which is an all-synchronous transmission facility. In this example, 300 bps asynchronous data sources are multiplexed into a 9.6 kbps synchronous data source for transmission onto DDS via the use of a DSU operating at 9.6 kbps. In the lower left portion of Figure 3.1, five 9.6 kbps asynchronous or synchronous data sources are multiplexed to obtain a 56 kbps synchronous data rate suitable for transmission on DDS. In both examples, one physical DDS circuit is used to transmit multiple logical channels of data.

The signals from the DSUs are terminated into a complementary office channel unit in the serving central office. From there, they enter into a multiplexing hierarchy which may carry voice as well as data.

Framing formats

One of the more interesting aspects of DDS is the constraints upon its transmission rate resulting from the formats used to encode user data. User data transmitted at 56 kbps is increased to a DS0 64 kbps data rate at the OCU, and that device inserts groups consisting of 7 bits of customer data into an 8-bit byte as illustrated in Figure 3.2A. In this encoding format the control bit (C) added to every 7 bits of customer data is set to a '1' if the byte contains customer data, while a value of '0' indicates that the byte contains network control data, such as idle or maintenance codes or control information. Since a DS0 signal results in the transmission of an 8-bit byte 8000 times per second, this framing format results in 8 kbps of control bits being added to the 56 kbps customer data rate.

The construction of DS0 signals from the 2.4, 4.8, and 9.6 kbps DDS subrates is illustrated in Figure 3.2B. As indicated, customer data is inserted into 8-bit bytes with 6 bits of user data framed by a frame bit (F) and a control bit (C). Once 2.4, 4.8, or 9.6 kbps DDS data streams are framed, one of two methods is used to place the framed data onto a DS0 channel. When 'byte stuffing' is used, the frame bit is set to '1' and the customer data is repeated the required

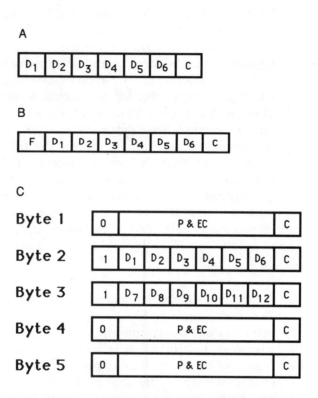

Figure 3.2 DDS framing formats: A 56 kbps: D = data bits, C = control bit; B 2.4, 4.8, 9.6 kbps: F = frame bit, D = data bits, C = control bit; C 19.2 kbps: D = data bits, C = control bit, P & EC = parity and error correction bits

number of times to create a 64 kbps DS0 signal. Thus, the 8-bit byte containing 6 bits of user data is repeated at 5, 10, and 20 times to enable 9.6, 4.8, and 2.4 kbps DDS data to be placed on a 64 kbps channel. When the F bit is set to '1' the frame format illustrated in Figure 3.2B, is referred to as a DS0-A format. Thus, DS0-A data can bypass the first level TDM illustrated in Figure 3.1 and be fed directly into the second level TDM.

The second method of placing 2.4, 4.8, or 9.6 kbps DDS data onto a DS0 channel involves the use of the first level TDM illustrated in Figure 3.1. When this occurs, five 9.6, ten 4.8, or twenty 2.4 kbps formatted signals are multiplexed onto a single DS0 channel. To distinguish between the repeating of the same data resulting from byte stuffing and the multiplexing of different DDS signals, the framing bit is altered from all ones in byte stuffing to a subrate framing pattern to indicate multiplexing of different DDS data sources. When this framing pattern occurs, the resulting framing format is referred to as a DS0-B format. Obviously, DS0-B formatting is more efficient than DS0-A formatting as the latter would require 20 DS0 channels to transmit twenty 2.4 kbps signals, while the former would require only one DS0 channel.

The introduction of 19.2 kbps DDS service required a substantial framing format change to accommodate this data rate. As illustrated in Figure 3.2C, 5 bytes are required to carry 19.2 kbps customer data since 3 bytes are used for parity and error correction functions. In this framing format the frame bit in each byte (bit 1) results in a '01100' repeating pattern. Twelve bits of customer data are placed into two 6-bit groups contained in bytes 2 and 3, resulting in 12 data bits being carried in every 5-byte group of 40 bits. Thus, the use of the 64 kbps DS0 channel produces an effective data rate of $64 \times 12/40$, or 19.2 kbps.

Signaling structure

A modified bipolar signaling structure is used on DDS facilities. The modification to bipolar return to zero signaling results in the insertion of zero suppression codes to maintain synchronization whenever a string of six or more zeros is encountered. Otherwise, repeaters on the span line routed between the carrier office and the customer may not be able to obtain clocking from the signal and could then lose synchronization with the signal.

To insure a minimum ones density, at 2.4, 4.8, 9.6, and 19.2 kbps any sequence of six consecutive zeros is encoded as 000X0V, where

0 denotes zero voltage transmitted (binary 0),
X denotes a zero or + or − A volts, with the polarity determined by conventional bipolar coding,
V denotes + or − A volts, with the polarity in violation of the bipolar rule.

Figure 3.3 illustrates the zero suppression sequence used to suppress a string of six consecutive zeros. For transmission at 56 kbps, any sequence of seven consecutive zeros is encoded as 0000X0V.

Timing

Precise synchronization is the key to the success of an all-digital

Format: 000X0V

Utilization

last binary one negative

000000 encoded as:

last binary one positive

000000 encoded as:

Figure 3.3 DDS zero suppression sequence

network. Timing insures that data bits are generated at precise intervals, interleaved in time and read out at the receiving end at the same interval to prevent the loss or garbling of data.

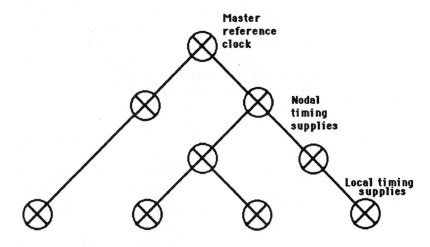

Figure 3.4 DDS timing subsystem

To accomplish the necessary clock synchronization on the AT&T digital network, a master clock is used to supply a hierarchy of timing in the network. Should a link to the master clock fail, the nodal timing supplies can operate independently for up to 2 weeks without excessive slippage during outages. In Figure 3.4, the hierarchy of timing supplies as linked to AT&T's master reference clock is illustrated. As shown, the subsystem is a treelike network containing no closed loops.

Service units

When DDS was introduced, both a channel service unit (CSU) and a data service unit (DSU) were required to terminate a DDS line.

The DSU converts the signal from data terminal equipment into the bipolar format used with DDS and T1 facilities.

The DDS master reference clock is an atomic clock that is accurate to 0.01 part per million (PPM). This clock was installed by AT&T at Hillsboro, MO, which is the geographic center of the United States and whose location insures a minimal variance in propagation delay time between DDS nodes connected to the master reference clock. The master reference clock oscillates at a rate known as the basic system reference frequency and is the most accurate of three timing sources used by digital facilities. The other two sources of timing include channel banks and loop timing where clocking is obtained from a high-speed circuit. A discussion of channel bank and loop timing is contained in Chapter 5.

The DSU interface to the DTE is accomplished by the use of a standard 25 pin EIA RS-232/V.24 female connector on the 2.4 kbps through 19.2 kbps units. The wideband, 56 kbps device utilizes a 34 pin CCITT, V.35 (Winchester) female type connector. Table 3.2 lists the RS-232 and V.35 interchange circuits commonly used by most DSUs. Since DTEs are normally attached to the DSU, the latter's interface is normally configured as data communications equipment (DCE) by the manufacturer.

Prior to deregulation, the CSU was provided by the communications carrier, while the DSU could be obtained from the carrier or from third-party sources. This resulted in an end-user connection to the DDS network similar to that illustrated in the top portion of Figure 3.5 where the CSU terminated the carrier's four-wire loop and the DSU was cabled to the CSU. In this configuration the CSU terminates the carrier's circuit. In addition, a separate CSU was designed to perform signal regeneration, monitor incoming signals to detect bipolar violations and perform remote loopback testing. Interfacing between the DSU and CSU is accomplished by the use of a 15-pin female D-type connector which utilizes the first six pins: where pin 1 is signal ground, pin 2 is status indicator, pins 3 and 4 are the receive signal pair, while pins 5 and 6 are the transmit signal pair. In addition to the communications carriers, several independent vendors now offer compatible CSU for customer interconnection to digital networks.

Table 3.2 DSU interchange circuits.

RS-232 Interface (DCE)		V.35 Interface (DCE)	
Pin	Signal	Pin	Signal
1	Chassis ground	P	Transmit data (A)
2	Transmit data	S	Transmit data (B)
3	Receive data	R	Receive data (A)
4	Request-to-send	T	Receive data (B)
5	Clear-to-send	C	Request-to-send
6	Data set ready	D	Clear-to-send
7	Signal ground	H	Data terminal ready
8	Carrier detect	E	Data set ready
9	Positive voltage	B	Signal ground
10	Negative voltage	F	Receive line signal detect
15	Transmit clock	Y	Transmit timing (A)
17	Receive clock	AA	Transmit timing (B)
20	Data terminal ready	V	Receive timing (A)
24	External transmit clock	X	Receive timing (B)
		U	External transmit timing (B)
		W	External transmit timing (B)
		L	Local loop-back

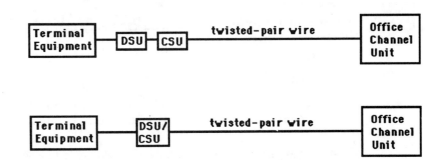

Figure 3.5 DSU/CSU connection

Since deregulation, over 20 third-party vendors have manu-factured combined DSU/CSU devices, integrating the functions of both devices into a common housing which is powered by a common power supply. The lower portion of Figure 3.5 illustrates the connection of end-user terminal equipment to DDS using a combined DSU/CSU unit.

DSU/CSU tests and indicators

Through the use of intentional bipolar violations, the DSU/CSU can generate a request to the OCU for the loop-back of the received signal onto the transmit circuit or it can interpret DDS network codes and illuminate relevant indicators on the device. When the loop-back button on the DSU/CSU is pressed, the device will generate four successive repetitions of the sequence 0B0X0V when operating at data rates up to 19.2 kbps or N0B0X0V at 56 kbps, where

B denotes + or − A volts, with the polarity determined by bipolar coding for a binary 1,

X denotes a zero volt for coding of binary 0 or B, depending upon the required polarity of a bipolar violation,

V denotes + or − A volts, with the polarity determined by the coding of a bipolar violation,

N denotes a don't care condition where the coding for a binary 0 or binary 1 is acceptable.

In Figure 3.6, the DDS loop-back 6-bit sequence is illustrated for data rates at or under 19.2 kbps. Note that the sequence transmitted is dependent upon whether the previous binary 1 was transmitted as a positive or negative voltage.

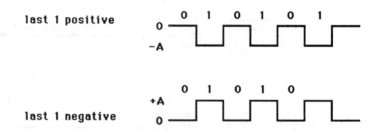

Figure 3.6 DDS loop-back seqence (data rates up to 19.2 kbps). DDS loop-back codes are intentional bipolar violations

Other bipolar violation sequences used by DDS include an idle sequence which indicates that a DTE does not have data to transmit, an out-of-service sequence and an out-of-frame sequence. The idle sequence is generated by the DSU while the out-of-service and out-of-frame sequences indicate a problem in the DDS network and are generated by the network and used by the DSU to illuminate an appropriate indicator on the device.

DDS II

During 1988, AT&T introduced a new version of its DDS facility commonly referred to as DDS II. One of the key advantages of DDS II is its capability to provide a diagnostic channel along with the primary subrate channel. This diagnostic channel is obtained through a modification to the framing used by DDS and requires the use of a special DSU/CSU that supports the new channel.

Through the use of DDS II, end-users can perform non-disruptive testing or use the channel for network management purposes. The key to obtaining the ability to derive a secondary channel on DDS is the use of the network control or C bit.

In conventional DDS the C bit, which is bit eight in each DDS 8-bit byte, is transmitted as a binary one whenever a DTE requests access to a channel by turning its request-to-send (RTS) signal on. With the C bit continuously set to a one the DTE can transmit an unrestricted stream of data to include continuous zeros since every eighth bit will be automatically set to a one. By robbing this bit once every third byte, AT&T established a virtual path for diagnostic use.

The diagnostic channel data rate for 56 kbps DDS II is obtained by multiplying the full DS0 rate of 64 kbps by $\frac{1}{8}$ which represents the C bit's portion of the DS0 rate to obtain 8 kbps. Next, since the bit robbing occurs every third byte, the resulting data rate becomes $8000 \times \frac{1}{3}$, or $2666\frac{2}{3}$ bps. Similarly, dividing $2666\frac{2}{3}$ bps by the number of 19.2, 9.6, 4.8, or 2.4 kbps channels multiplexed onto a DS0 channel results in the diagnostic data rate for DDS II at those data rates.

Currently, AT&T is adding diagnostic channel capability throughout the DDS network. Although older DSU/CSU devices can support transmission on DDS II, those devices cannot support the use of the secondary channel capability provided by this modification to DDS. To do so requires the use of newer DSU/CSU devices that support the multiplexing of diagnostic data onto every third C bit position.

Analog extensions to DDS

AT&T provides an 831A auxiliary set which allows analog access to DDS for customers located outside the DDS servicing areas. The 831A connects the EIA RS-232 interfaces between a data service unit and a modem. The 831A contains an 8-bit elastic store, control, timing, and test circuits which allow loop-back tests toward the digital network. The elastic store is a data buffer that is required by the DSU to receive data from the modem in time with the modem's receive clock. The data is then held in the elastic store until the DSU's transmit clock requests it. Thus, the buffer serves as a mechanism to overcome the timing differences between the clocks of the two devices. In the reverse direction, no buffer is required when the DSU's receive clock is used as the modem's external transmit clock. When the modem cannot be externally clocked or when one DSU is connected to a second DSU or a DTE that cannot accept an external clock, a second elastic store will be required. Figure 3.7 illustrates a typical analog extension to a DDS servicing area.

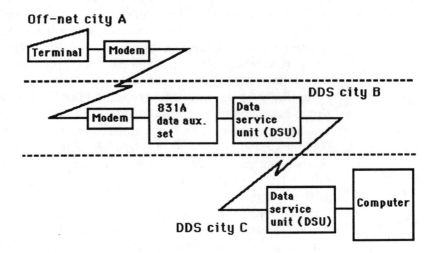

Figure 3.7 Analog extension to DDS. An analog extension to DDS requires the use of one or more elastic stores to compensate for timing differences between a modem and a DSU

3.1.2 British Telecom KiloStream

KiloStream is a point-to-point leased line digital service that was first offered commercially in January, 1983. Although KiloStream is similar to DDS, there are several significant differences that warrant discussion.

Like DDS, KiloStream is an all-synchronous facility. British Telecom provides a network terminating unit (NTU) which is similar

to DSU/CSU to terminate the subscriber's line. The NTU provides
a CCITT interface for customer data at 2.4, 4.8, 9.6, or 48 kbps to
include performing data control and supervision, which is known
as structured data. At 64 kbps, the NTU provides a CCITT interface
for customer data without performing data control and supervision,
which is known as unstructured data.

The NTU controls the interface via CCITT recommendation X.21,
which is the standard interface for synchronous operation on
public data networks. An optional V.24 interface is available at
2.4, 4.8, and 9.6 kbps, while an optional V.35 interface can be
obtained at 48 kbps. The X.21 interface is illustrated in Figure 3.8.
Here the control circuit (C) indicates the status of the transmitted
information—data or signaling, while the indication circuit (I)
signals the status of information received from the line. The control
and indication circuits control or check the status bit of an 8-bit
envelope used to frame six information bits.

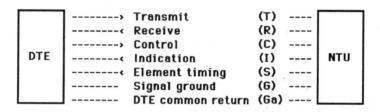

Figure 3.8   CCITT X.21 interface circuits

Data encoding

Customer data is placed into a 6+2 format to provide the signaling
and control information required by the network for maintenance
assistance. This is known as envelope encoding and is illustrated
in Figure 3.9.

The NTU performs signal conversion, changing unipolar non-
return to zero signals from the V.21 interface into a di-phase WAL
2 encoding format. This insures that there is no dc content in the
signal transmitted to the line, provides isolation of the electronic
circuitry from the line, and provides transitions in the line signal
to enable timing to be recovered at the distant end. Table 3.3 lists
the NTU operational characteristics of KiloStream.

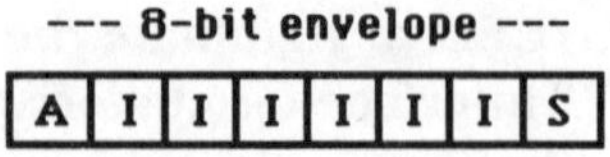

--- 8-bit envelope ---

A	I	I	I	I	I	I	S

Figure 3.9 KiloStream envelope encoding: A = alignment bit which alternates between '1' and '0' in successive envelopes to indicate the start and stop of each 8-bit envelope, S = status bit which is set or reset by the control circuit and checked by the indicator circuit, I = information bits

Table 3.3 KiloStream NTU operational characteristics.

Customer data rate (kbps)	DTE/NTU interface	Line data rate (kbps)	NTU Operation
2.4	X.21	12.8	6 + 2 envelope encoding
4.8	X.21	12.8	6 + 2 envelope encoding
9.6	X.21	12.8	6 + 2 envelope encoding
48	X.21	64	6 + 2 envelope encoding
64	X.21	64	No envelope encoding
48	X.21 bis/V.35	64	6 + 2 envelope encoding
2.4	X.21/V.24	12.8	6 + 2 envelope encoding
4.8	X.21/V.24	12.8	6 + 2 envelope encoding
9.6	X.21/V.24	12.8	6 + 2 envelope encoding

The KiloStream network

In the KiloStream network, the NTUs on a customer's premises are routed via a digital local line to a multiplexer operating at 2.048 Mbps. This data rate is the European equivalent of the T1 line in the United States that operates at 1.544 Mbps. The multiplexer can support up to 31 data sources and may be located at the local telephone exchange or on the customer's premises if traffic justifies. It is connected via a digital line or a radio system into the British Telecom KiloStream network as illustrated in Figure 3.10.

Unlike true CEPT, the 2.048 Mbps T-carrier used for KiloStream uses 31 DS0 channels. Normally, CEPT uses one channel for synchronization (framing) and a second channel for signaling. Since there is no direct voice signaling on KiloStream, time slot 16, which normally would carry that information, can be used for data. From an examination of Figure 3.10, the reader will note that a DTE operating rate of 2.4, 4.8, or 9.6 kbps results in a line rate of 12.8 kbps, which is precisely one-fifth of the DS0 64 kbps data rate.

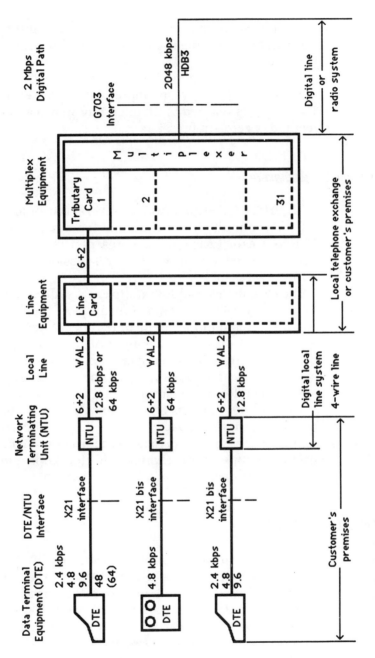

Figure 3.10 KiloStream structure

Thus, multiplexers used by British Telecom at their local exchange are capable of placing five low-speed KiloStream circuits onto each DS0 channel.

The HDB3 line coding shown in Figure 3.10 represents the CCITT method used to insure an appropriate ones density on a T-carrier. The reader is referred to Chapter 4 for information concerning this method of zero suppression.

Representative multiplexers

Both British Telecom as well as several third-party vendors market a variety of multiplexers to extend the functionality and capability of KiloStream usage. Two representative products marketed by British Telecom that warrant discussion are the K3 and K5 multiplexers.

The British Telecom K3 multiplexer enables the KiloStream user to obtain a voice and data transmission capability on a 64 kbps channel. This multiplexer includes one 32 kbps CVSD modulation adapter that digitizes voice input at 32 kbps. In addition to voice digitization, the K3 also supports up to two synchronous channels at data rates up to 9.6 kbps and one data source at 19.2 kbps on a 64 kbps KiloStream circuit.

The British Telecom K5 multiplexer can be used to maximize the data-carrying capacity of a 64 kbps KiloStream circuit. This multiplexer supports up to five synchronous channels at data rates up to 9.6 kbps or two channels at 19.2 kbps and one channel at 9.6 kbps for operation on a 64 kbps circuit.

3.2 THE T-CARRIER

Due to the superior quality provided by T-carrier facilities for supporting voice and data transmission, they are now routinely available for commercial use in most parts of North America, Europe, and Japan. In this section we will examine the physical construction of the T-carrier facility from the carrier's serving office to an organization's premises. In addition, due to the divestiture by AT&T of its operating companies, we will also focus our attention upon the pricing components related to the installation and operation of a T1 circuit that crosses local access transport area (LATA) boundaries in the United States. Using the preceding information as a base, we will then examine how these high-speed transmission facilities are typically used by end-user organizations.

3.2.1 T-carrier line structure

Unlike the interface for DDS that requires the use of a combined DSU/CSU or separate devices, only a CSU is required to terminate a T-carrier in the United States. The removal of the DSU results from the incorporation of its functionality into T1 multiplexers and channel banks.

Figure 3.11 illustrates the T1 line structure routing from a communications carrier's nearest serving office to the customer premises. The first repeater located outside of the serving central office is typically located 4500 feet from that office. Thereafter, T1 repeaters are typically spaced 6000 feet apart from one another.

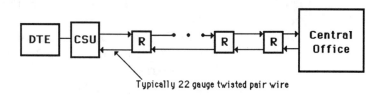

Typically 22 gauge twisted pair wire

Figure 3.11 T1 line structure: CSU = channel service unit, R = repeater, DTE = data terminal equipment

Since repeaters are active devices that regenerate digital pulses, they require power. Power for repeaters is provided by the communications carrier which uses a power supply at the serving central office to place a voltage between the tip and ring leads on the circuit. Both the voltage and current vary dependent upon the length of the line to the customer premises and the gauge of the wire. The most common voltage used is a nominal −48 volt DC source, resulting in a −60 mA current on a low-power T1 circuit and a −140 mA current on a standard T1 circuit. This type of powered circuit, which is completed by the CSU, is commonly referred to as a 'Wet T1'. A second type of T1 line provides a data transportation path through the use of optical fiber that transports light energy instead of electrical energy. When an optical fiber line is used to connect the customer premises to the serving central office, the facility is referred to as a 'Dry T1' line due to the absence of power.

Since T1 CSUs were originally constructed to obtain power from the T1 line, equipment that is not capable of being locally powered cannot be used with a dry T1 line.

3.2.2 T1 cost components

Prior to 1984 when AT&T's divestiture became effective, there were only two types of tariffs–interstate and intrastate. Interstate tariffs were filed by AT&T and other long-distance carriers with the Federal Communications Commission, while intrastate tariffs were filed by local operating companies with state public utility commissions.

Although the distinction between interstate and intrastate communications is still used to determine the applicable regulatory body, a new criteria with respect to line costs resulted from divestiture. This criteria concerns whether or not a service is within the local area served by the divested local operating companies. These areas are known as local access and transport areas, or LATAs, and approximately correspond to the standard metropolitan statistical areas defined by the US Commerce Department.

If service provided by a communications carrier links two locations within a LATA, it is an inter-LATA service. If a service must cross LATA boundaries, the local exchange carrier (LEC) must connect their facility to an inter-LATA carrier, also known as an interexchange carrier (IEC). Both IECs and LECs have tariffs. These tariffs govern the use of IEC facilities and the use of local facilities by the IEC, the latter being the method by which LECs charge IECs for the use of their local access facilities.

Based upon the preceding, end-users may have to contend with up to six types of tariffs: inter-LATA, intra-LATA, and LATA access, for both interstate and intrastate communications. Within each LATA is a series of interface points known as points of presence (POP). Each IEC, such as AT&T, MCI, and US Sprint, has its own POPs which are the only locations within a LATA where the IEC can receive and deliver traffic. Due to this structure, a customer's premises must use the facilities of the LEC to connect to the IEC's network at the latter's POP.

In our examination of the cost elements of a T1 facility we will focus our attention upon an inter-LATA circuit. This type of circuit will consist of three components:

- a local circuit from the end-user's near end premises to the POP of their selected IEC,

- an interoffice circuit that connects the near end IEC's POP to the far end IEC's POP,

• a local circuit from the IEC's far end POP to the end-user's far end premises.

Figure 3.12 illustrates an example of the routing structure of a T1 circuit between Macon, Georgia, and Washington, DC. Note that the local access channel in Macon, Georgia, is normally provided by Southern Bell, while the local access channel in Washington, DC, would be provided by Bell Atlantic. To minimize the cost associated with local access channels, many organizations are using the facilities offered by cable TV (CATV), fiber optic local transmission systems, and microwave. Collectively, these techniques are known as bypass access since they bypass the LEC and route the user's organization to the IEC point of presence.

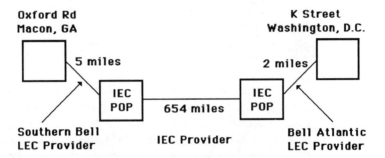

Figure 3.12 Routing structure of a T1 circuit: LEC = local exchange carrier, IEC = interexchange carrier, POP = point of presence

Although access to the IEC normally occurs through the use of LEC facilities, many organizations also employ bypass facilities to access the IEC point of presence. In many instances, the IEC point of presence may be in the same building occupied by the LEC. Thus, the connection between the two may be no more than a short cable connecting the equipment of the two communication carriers. Although most IECs have a POP in each LATA, the reader should note that this does not insure that all IEC facilities are available for connection. Thus, some AT&T POPs may not support DDS or T1 connections.

For illustrative purposes, Table 3.4 lists the recurring (monthly) and one-time costs (installation) associated with the T1 circuit shown in Figure 3.12. The entries in this table are approximations used for illustrative purposes and, while they were relatively accurate at the time this book was published, the reader should note that they may have been rendered obsolete due to tariff changes.

Table 3.4 T1 circuit cost components.

	Monthly ($)	Installation ($)
Local access channel		
Macon, GA		
Fixed	691	2 227
Mileage 5 at 40	200	
Central office connection	62	310
Access coordination	22	207
Interchange channel		
Fixed	2 600	
Mileage 650 at 15.50	10 075	
Local access channel		
Washington, DC		
Fixed	579	1 188
Mileage 2 at 45	90	
Central office connection	62	310
Access coordination	22	207
Total	14 403	4 449

The local access channel costs, in many instances, are very disproportionate to the circuit mileage. Due to this, many end-user organizations have employed the use of third-party carriers to obtain access to the IEC POP via coaxial cable, microwave, or fiber optic transmission. As previously discussed, this method of accessing the POP is known as bypass access.

The central office connection and access coordination fees are normally billed by the IEC. The central office connection fee is the cost for the IEC connecting the local channel to an interoffice channel. The access coordination fee covers the cost of the IEC providing a single point of contact by coordinating installation and testing with the LEC.

3.2.3 Digital access and cross connect

In 1981 AT&T introduced its digital access and cross connect system (DACS) to facilitate testing as well as to reduce maintenance costs. Through the use of DACS, DS0 time slots can be dropped from one T1 circuit and added to another T1 line, a process known as drop and insert.

DACS cross connections are administered via software control. Originally this control was restricted to initiation from a communications carrier central office. Later, the ability to control DACS cross connections was extended to end-user locations. The latter capability is known as customer controlled reconfiguration (CCR) and is a service function of AT&T's Accunet™ 1.5 service, as well as a function available from other communications carriers.

DACS have the capability to switch both subrate and full rate T1 channels. For internal networks, one vendor now markets an IBM PC AT personal computer with two add-in boards that can terminate four T1 circuits and switch all of the DS0 subchannels to or from each T1 line. With another vendor's PC AT product, internal corporate network users can terminate up to 16 T1 lines and manage up to 384 individual DS0 subchannels. No matter where performed or the type of computer control used, DACS perform certain basic functions to include drop/insert/bypass and groom and fill operations.

Drop/insert/bypass

The drop/insert/bypass capability is used to satisfy the requirement for some voice and data channels to terminate at a node, while other channels must bypass a node and are then routed on another circuit towards another node.

When a channel is dropped at a node or bypasses a node and is routed onto a different circuit, its drop or removal results in wasted capacity. This capacity can be filled by inserting other channels that have the same intermediate or final destination as the channel that bypassed a node. Figure 3.13A illustrates a DACS bypass operation.

Groom and fill operations

In addition to drop/insert/bypass, some DACS perform a groom and fill operation on rerouted data to maximize the data handling efficiency of the facility. The groom function is used to segregate channels for transmission to their appropriate end points as illustrated in Figure 3.13B. The fill function is often used with the groom function as the former maximizes the efficiency of a large number of groom operations. Here the fill feature attempts to combine traffic from two or more carriers onto one carrier with the same route destination. This feature, which is illustrated in of Figure 3.13C, as well as the groom function, is usually incorporated into large DACS systems that control carrier facilities.

One example of the use of DACS within a carrier's communications network is shown in Figure 3.14, which illustrates British Telecom's KiloStream network routing. In this example, note that switching occurs on 64 kbps DS0 channels, which explains why

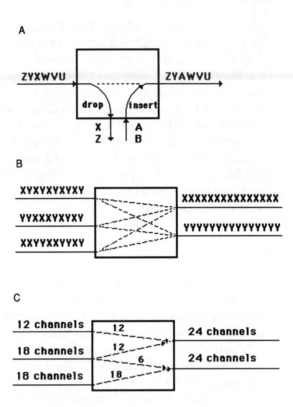

Figure 3.13 DACS functions: A drop/insert/bypass; B groom operation; C fill operation

low-speed 2.4 through 9.6 kbps data streams are operated at a line speed of 12.8 kbps on KiloStream, since five of those data streams can then be carried within a switched DS0 channel.

3.3 ACCUNET T1.5 AND MEGASTREAM UTILIZATION

Both Accunet T1.5 offered by AT&T and MegaStream offered by British Telecom are T-carrier transmission facilities for high-speed data and high-volume speech communications.

The typical utilization of each T-carrier is a transportation facility for data sources that can be effectively serviced by the high data rate provided by a T-carrier. Such usage includes the servicing of digital PABXs, analog PABXs with voice digitizers, clustered terminals via multiplexer input analog terminations, data circuits, computer-to-computer transmission, and video conferencing.

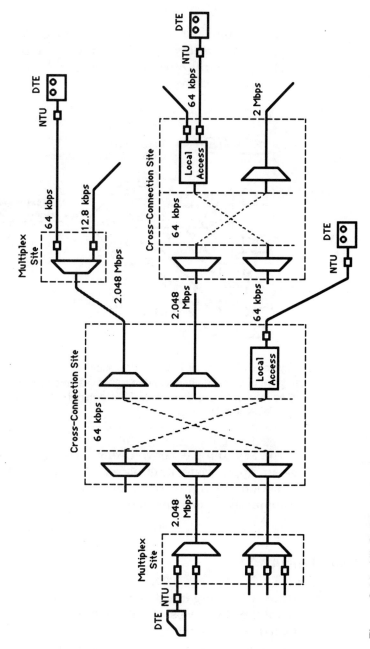

Figure 3.14 KiloStream network routing

The Accunet T1.5 facility operates at 1.544 Mbps, while the MegaStream facility operates at the CEPT 2.048 Mbps data rate.

The typical utilization of an Accunet T1.5/MegaStream facility is illustrated in Figure 3.15.

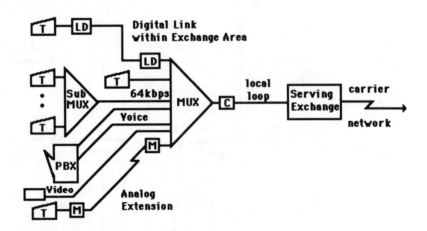

Figure 3.15 Typical T-carrier utilization. Through the use of T-carrier facilities a mixture of voice, data, and video transmission can be routed over a common high speed circuit: LD = line driver, T = terminal, M = modem, C = channel service unit, MUX = multiplexer.

The key to the effective use of T-carrier facilities is the selection of an appropriate multiplexer. In Chapter 6 we will focus our attention upon the operational capabilities of this category of communications equipment.

3.4 FRACTIONAL T1

Until 1989, organizations that required a data transportation capability in excess of that offered on DDS or KiloStream were forced to either migrate to a T-carrier or obtain multiple subrate circuits. In migrating to a T-carrier the end-user had to pay for twenty-four or thirty 64 kbps channels on a four-wire circuit regardless of the number of channels they actually needed. If multiple subrate circuits were used, the cost associated with the central office connection and coordination fees, as well as multiple DTEs, such as multiplexers, usually made that solution uneconomical.

Based upon the preceding, AT&T was one of several communications carriers to introduce a fractional T1 service. Marketed under the name Accunet Spectrum of Digital Services (ASDS), users of this service can lease a fractional portion of the T1 bandwidth in 64 kbps DS0 increments. ASDS provides digital interoffice channels at 56 or 64 kbps and intermediate bit rates of 128, 256, 384, and 768 kbps.

The benefits of fractional T1 are considerable. In addition to saving the cost of leasing an entire T1 line when only a portion of the bandwidth is required, this facility also provides a method for orderly growth. Thus, an organization could grow from 128 kbps today to 256 kbps tomorrow. Another significant advantage of ASDS is its announcement cost in comparison to both 56 kbps DDS and analog voice-grade line tariffs. On a per mile basis, the cost of a 64 kbps FT1 facility was the same as an analog line and approximately one-third to one-fifth the cost of a 56 kbps DDS circuit. Although AT&T's initial deployment of ASDS was limited to 25 points of presence (POP) in late 1989, by 1991 this service was scheduled to become available on a nationwide basis.

Access to FT1 is similar to the method illustrated in Figure 3.12 for access to a T1 circuit. Although, at the time this book was written, all LECs provided access to FT1 by using a full T1 circuit that is interconnected at the POP at fractional T1 rates, it is expected that LECs will eventually offer FT1 service to the IEC FT1 facilities.

3.5 DS3 OPERATION AND UTILIZATION

Although fractional T1 satisfies the user who does not require a full T1 circuit, it did not alleviate the problem large organizations faced when their data and voice transmission requirement began to occupy multiple T1 circuits. To provide assistance to large users in meeting their demand for higher capacity transmission, several carriers have made their DS3 facilities available for commercial use.

The DS3 signal operates at a rate of 44.736 Mbps and carries 28 T1 signals, which is the equivalent of 672 voice channels. Although it may be quite some time before most organizations have to consider the use of this facility, its commercial availability provides a migration path that should satisfy the voice, data, and video transmission requirements of just about every commercial organization and governmental agency.

REVIEW QUESTIONS

1 What equipment would an organization use to transmit several asynchronous data sources over one synchronous DDS network facility?

2 What is the purpose of the control (C) bit in a DDS byte?

3 Compare the efficiency of DDS byte stuffing to DDS multiplexing. Which is more efficient?

4 Illustrate how a string of seven consecutive zeros would be encoded on DDS at a transmission rate of 56 kbps if the last binary one was encoded as a positive pulse.

5 Illustrate the encoded sequence a DSU/CSU operating at 56 kbps would generate when the loop-back button is pressed when the device operates at 56 kbps.

6 What is the diagnostic channel rate of a 2.4 kbps DDS II transmission facility?

7 Explain how you could use a 56 or 64 kbps DDS or KiloStream facility to carry voice and data.

8 What is the difference between a 'wet' and a 'dry' T1 line?

9 What is a point of presence?

10 What is the primary rationale behind bypassing a local exchange carrier to access the point of presence of an interexchange carrier?

11 How could you use a digital access cross connect system at location B to route three DS0 channels from location A to C via location B if T1 lines connect A to B and B to C?

4

T-CARRIER FRAMING
AND CODING
FORMATS

In this chapter we will focus our attention upon both North American and European T-carrier framing and coding formats. Our examination of T-carrier framing will include an investigation of the different types of framing used, as well as the method by which mechanical signaling and performance information is transported, and how the frame structure is designed to provide synchronization and report alarm conditions. Since maintaining a minimum number of binary ones on a T-carrier is critical for repeater operations, we will conclude this chapter by examining several coding formats that are employed to insure a minimum ones density on a T-carrier circuit.

4.1 T-CARRIER FRAMING

Since the framing structures used on North American and European T-carrier facilities vary considerably from one another, we will examine each structure as a separate entity in this section.

4.1.1 North America

In North America the T-carrier was designed to transmit 24 independent voice channels with each channel encoded as a 64 kbps data stream.

D4 framing

In North America the T1 signal represents a composite of 24 separate DS0 channels, each representing one PCM encoded voice signal digitized at 64 kbps. The 24 channels in a T1 signal are multiplexed in a round-robin order to insure each channel is transmitted in turn and that every channel receives a turn prior to any channel receiving a second turn. To denote the beginning of each sequence of 24 digitized DS0 channels, a special bit called the frame bit is prefixed to the beginning of each multiplexing cycle. Since each DS0 channel is encoded into a PCM word using 8 data bits, 24 DS0 channels represents a sequence of 192 data bits. The full pattern of 1 frame bit and 192 data bits is called the DS1 (digital signal, level 1) frame and represents a total of 193 bits. Since sampling occurs 8000 times per second, 193 bits × 8000 samples per second results in the 1.544 Mbps operating rate of a T1 circuit. As this operating rate includes 8000 frame bits, only the remaining 1.536 Mbps is actually available to the user.

One of the most popular methods of framing the DS1 signal is called D4 framing. This framing technique takes its name from the AT&T D4 channel bank used in that communications carrier's network. Under D4 framing the frame bits in 12 consecutive frames are grouped together to form a superframe whose frame bits are used to form a repeating pattern. Figure 4.1 illustrates the D4 framing structure and framing pattern. Under D4 framing, the

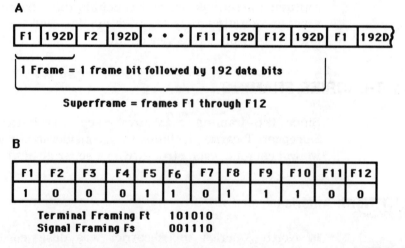

Figure 4.1 D4 framing structure and framing pattern. The D4 framing pattern represents the hex characters 8CD which are continuously repeated: A D4 framing structure; B D4 framing pattern

1.544 Mbps data stream must meet the following requirements.

- It must be encoded as a bipolar, AMI, non-return to zero signal to insure that the signal has no dc component and can be transformer coupled, permitting the circuit to carry power for the repeaters.

- Each pulse must have a 50 percent duty cycle with a nominal voltage of 3.0 volts.

- There can be no more than 15 consecutive '0s' present in the data stream, which defines the minimum ones density of the circuit.

- The D4 framing pattern is embedded in the data stream.

The framing bits in the D4 superframe consist of six Ft (terminal framing) bits that are used to synchronize the bit stream and six Fs (signal framing) bits that are used to define multiframe boundaries as well as to identify what is known as robbed bit signaling in frames 6 and 12. Robbed bit signaling is discussed later in this chapter.

The Ft bit conveys a pattern of alternating 0s and 1s (101010), which is used to define frame boundaries, enabling one slot to be distinguished from another. Due to this, it is also known as a frame alignment signal. The Fs bit conveys a pattern of 001110, which is used to define multiframe boundaries. This enables one frame to be distinguished from another, permitting frames 6 and 12 to be identified for the extraction of their signaling bits. Note that the composite D4 framing pattern represents the hex characters 8CD which are continuously repeated.

Since D4 framing represents a constant framing pattern, it can be utilized to determine the approximate bit error rate of a T1 line by monitoring. This can be accomplished by the use of a test set that can monitor the framing bit error rate which will usually be reflective of the error rate on the line. Unfortunately, for a more accurate measurement of the line error rate, transmission must be interrupted to permit the use of more sophisticated testing equipment.

Bit robbing

In voice transmissions, mechanical signaling information, such as 'on-hook' and 'off-hook' conditions, dialing digits, and call progress information, must be transmitted separately for each voice channel by including signaling information in the data stream. This is accomplished by robbing the eighth bit in every sixth and twelfth frame to transmit and receive voice channel signaling information.

The process of transmitting signaling data associated with each voice circuit within the voice channel is known as associative signaling. In comparison, the use of a common channel dedicated to carrying the signaling data for all voice circuits within a T-carrier link is called common channel signaling.

Under the D4 format, channels 1 through 5 and 7 through 11 use all bits for information coding. Since there are no special bits assigned for signaling, the least significant bit position of each channel in every sixth and twelfth frame is robbed for signaling.

Bits taken from the sixth frame are referred to as 'A' bits, while bits taken from the twelfth frame are referred to as 'B' bits.

Figure 4.2 illustrates the bit robbing process used to convey voice signaling information. Note that the use of the least significant bit in each DS0 PCM word in the sixth and twelfth frames minimizes the duration of bit robbing and its effect upon a voice conversation. This is because each DS0 PCM word has a duration of 125 μs, resulting in 625 μs of full 8-bit PCM words (125 × 5) for every 125 μs, where the word loses one bit of accuracy.

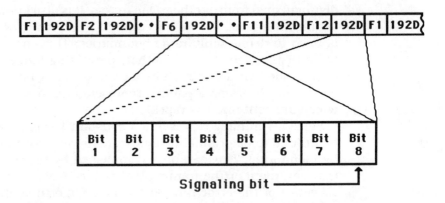

Figure 4.2 Bit robbing. Every eighth bit in the sixth and twelfth frames is used for signaling

Since only frames 6 and 12 contain associated signaling, those frames must be distinguishable from one another. This is the basis for the multiframe D4 structure consisting of 12 frames that are distinguished by the D4 framing pattern previously illustrated in Figure 4.1B.

One example of the use of robbed bit signaling is to pass E&M (ear and mouth) status information. Here the terms ear and mouth relate to the speaker and receiver in a telephone handset. Under robbed bit signaling, E&M status information is passed by varying the setting of the A and B bits. When the A bit is set to zero it denotes an active line, while it is set to a one when the line is inactive. The B bit is used to pass on- and off-hook information, with the B bit set to zero to indicate an off-hook condition, while it is set to a one to indicate an on-hook condition.

Extended superframe format

As previously discussed, D4 framing only provides an indirect measurement of line quality through the monitoring of frame bits. Another limitation of the D4 format is the fact that obtaining a communications capability between devices on a T1 circuit required the use of a DS0 time slot. To alleviate these problems, as well as to provide the T1 user with additional capability, AT&T introduced an extended superframe format in early 1985. Although this new framing format requires the installation of equipment that supports the frame format and has consequently been slowly introduced, within a few years a majority of T1 circuits in North America can be expected to conform to it. Eventually, the extended superframe format can be expected to replace D4 framing.

Denoted as Fe and ESF, the extended superframe format extends D4 framing to 24 consecutive frame bits—F1 through F24—as illustrated in Figure 4.3.

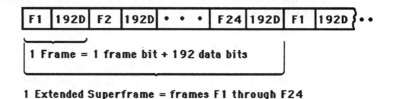

1 Extended Superframe = frames F1 through F24

Figure 4.3 The extended superframe

Unlike D4 framing in which the 12 framing bits form a specific repeating pattern, the ESF pattern can vary. ESF consists of three types of frame bits.

Derived data link

The ESF 'd' bits, which represent a derived data link, are used by the telephone company to perform such functions as network monitoring to include error performance, alarm generation and reconfiguration to be passed over a T1 link. The 'd' bits appear in the odd frame positions, e.g., 1, 3, ..., 21, 23. Since they are used by 12 of the 24 framing bits, the 'd' bits represent a 4 kbps data link.

The data link formed by the 12 'd' frame bits is coded into higher-level data link control (HDLC) protocol format known as BX.25. Figure 4.4 illustrates the data link format carried by the 'd' bits.

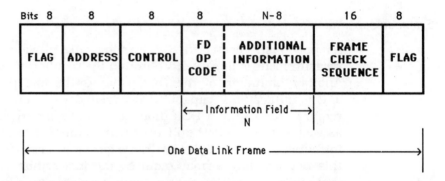

Figure 4.4 ESF data link format

The flag byte consists of the 8-bit sequence 01111110 and initiates and terminates each frame. The address field is used to identify a frame as either a command or response. A command frame contains the address of the device to which the command is being transmitted, while a response frame contains the address of the device sending the frame.

The control field identifies the purpose of the frame and can indicate one of three frame types—supervisory, unnumbered, or information. A supervisory frame is used for data link housekeeping information, such as acknowledgements. An unnumbered frame is used for major system commands, line initialization, and shutdown information, while an information frame contains user data. The frame check sequence (FCS) is a 16-bit CRC check used to insure the integrity of the data link.

One of the primary goals in the development of the BX.25 protocol was to provide a mechanism to extract performance information from ESF compatible CSUs. Doing so allows circuit quality monitoring without taking the circuit out of service and is a major advantage of ESF over D4 framing. Standard maintenance messages that are defined in AT&T's publication 54016 include messages that can return performance data concerning the number of errored seconds (ES), severely errored seconds (SES), and failed seconds (FS). An errored second is a second that contains one or more bit errors, while a severely errored second is considered to be a second with 320 or more bits in error. If ten consecutive severely errored seconds occur, this condition is considered as a failed signal state. Then each signal in a failed signal state is considered to be a failed second. Table 4.1 summarizes the standard maintenance messages transmitted on the ESF data link.

Table 4.1   ESF data link maintenance messages.

Send one-hour performance data
Upon receiving this command the ESF CSU will supply the following:
current status, elapsed time of current interval,
ES and FS in the current 15-minute interval,
number of valid intervals, count of ES and FS in 24-hour register,
ES and FS during the previous four 15-minute intervals.

Send 24-hour 'ES' performance data
Upon receiving this command, the ESF CSU will supply
the following: current status, elapsed time of current interval,
ESs and FSs in current 15-minute interval,
number of valid intervals, count of ESs and FSs in 24-hour register,
and ESs during previous ninety-six 15-minute intervals.

Send 24-hour 'FS' performance data
Upon receiving this command, the ESF CSU will supply
the following: current status, elapsed time of current interval,
ESs and FSs in the current 15-minute interval,
number of valid intervals, count of ESs and FSs in 24-hour register,
and FSs during previous ninety-six 15-minute intervals.

Reset performance monitoring counters
Upon receiving this command, the ESF CSU will reset all
interval times and ES and FS registers and supply the
current status.

Send errored ESF data
Upon receiving this command, the ESF CSU will supply
current data present in ESF error event registers. Each
count represents one error event (65535 maximum).

Reset ESF register
Upon receiving this command, the ESF CSU will reset the
ESF error event register and supply the current status.

Maintenance loop-back (DLB)
Energizes upon receiving the proper code embedded in the
4 kbps data link. This loop-back loops through the entire CSU.

Error check link

Frame bits 2, 6, 10, 14, 18, and 22 are used for a CRC-6 code. The 6-bit cyclic redundancy check sum is used by the receiving equipment to measure the circuit's bit error rate and represents 2 kbps of the 8 kbps framing rate. The CRC employs a mathematical algorithm which is used to check all 4632 bits in the ESF. Mathematically, the CRC check bit generation is peformed by the use of a fixed polynomial whose composition is $X_6 + X + 1$, or 1000011. The data block is first multiplied by X_6 or 1000000 and then divided by the polynomial. The remainder is then transmitted in the six ESF CRC bit positions. At the receiver, a similar operation is performed on the received data block using the same polynomial and the locally generated check bit sequence is then compared to the received check bit sequence. The CRC-6 code yields an accuracy of 98.4 percent, and the occurrence of a mismatch between the locally generated check bit sequence and the received check bit sequence indicates that one or more bits in the extended superframe are in error.

To conform with ESF CRC-6 coding and reporting requirements, the use of an ESF compatible channel service unit is required. This CSU not only generates the CRC-6 but must also be capable of detecting CRC errors and storing a CRC error count over a 24-hour period. ESF compatible CSUs contain buffer storage which enables the device to store current line status information, including all error events and errored and failed seconds for the current 15-minute period and the previous ninety-six 15-minute periods that represent the prior 24-hour period. To enable the carrier to retrieve this data, as well as to reset any or all counters and activate or deactivate loop-back testing on the local span line, the CSU must also have the ability to respond to network commands. Thus, an ESF compatible CSU must have the capability to send and receive data based upon the BX.25 formation via the 4 kbps data link.

A competing standard from the T1E1 committee requires the CSU to broadcast this information all of the time instead of upon request from the carrier. This standard is designed to overcome the potential problem resulting from an IEC resetting CSU registers prior to an LEC reading the CSU registers.

Framing pattern

The third type of frame bits are used to generate the framing pattern. Here, frame bits 4, 8, 12, 16, 20, and 24 are used to generate the Fe framing pattern whose composition is 001011.

These six bits result in a 2 kbps framing pattern. Another difference between the ESF and D4 frame format is in the area of signaling. ESF has added two additional signaling bits, C and D, in frames 18 and 24. Thus, in ESF signaling data is accommodated by using bit robbing in frame 6 (A-bit), frame 12 (B-bit), frame 18 (C-bit), and frame 24 (D-bit). In Table 4.2 the reader will find a summary of the ESF framing pattern.

Table 4.2 ESF framing pattern.

Frame	Bit composition	Frame	Bit composition
1	d	13	d
2	C1	14	C4
3	d	15	d
4	0	16	0
5	d	17	d
6	C2	18	C5
7	d	19	d
8	0	20	1
9	d	21	d
10	C3	22	C6
11	d	23	d
12	1	24	1

d = data link, Cx = CRC-6 bit x.

T1 alarms and error conditions

There are several alarms and error conditions that are monitored and reported under the T1 D4 and ESF formats. Principal T1 alarms include a red alarm which is produced by a receiver to indicate that it has lost frame alignment and a yellow alarm which is returned to a transmitting terminal to report a loss of frame alignment at the receiving terminal. Normally, a T1 terminal will use the receiver's red alarm to request that a yellow alarm be transmitted.

To illustrate the operation of T1 alarms, consider the configuration illustrated in Figure 4.5 that shows two channel banks connected together via the use of a four-wire T1 circuit. If the second channel bank (CB2) loses synchronization with the first channel bank (CB1) or completely loses the signal it then transmits a red alarm to CB1. The transmission of the red alarm is accomplished by CB2 forcing bit position 2 in each PCM word to zero and the frame signaling bit in frame 12 to a binary 1 under the D4 format. If ESF framing is used, CB2 sends a repeated pattern of eight zeros and eight ones on the data link.

When CB1 recognizes the red alarm, it transmits a yellow alarm. This alarm, in effect, says 'I'm sending data but the other end is not receiving the transmitted information and the problem is elsewhere.' Under D4 framing the yellow alarm is generated at the receiver by setting bit 2 to zero for 255 consecutive channels and the frame alignment signal (Fs) to one in frame 12. Under ESF a pattern of eight zeros and eight ones repeated 16 times is used to indicate a yellow alarm.

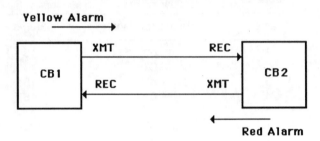

Figure 4.5 T1 alarm generation example. A red alarm is produced by a receiver to indicate it has lost frame alignment, while a yellow alarm is returned to a transmitting terminal to report a loss of frame alignment at the receiving terminal

A red alarm is generated when either the network or a DTE senses an error in the framing bits for either two out of four or five framing bits and this condition persists for more than 2.5 seconds. Table 4.3 summarizes the method of transmitting alarms on D4 and ESF T1 circuits.

Table 4.3 T1 alarm formats.

Mode	Format
Transmitted red alarm	
T1 D4	bit 2=0 in all data channels and Fs=1 in frame 12
T1 ESF	repeated pattern of 8 zeros, 8 ones on data link
Yellow alarm generated at receiver	
T1 D4	bit 2=0 for 255 consecutive channels and Fs=1 in frame 12
T1 ESF	16 patterns of 8 zeros, 8 ones on data link

In addition to red and yellow alarms, a third type of alarm that warrants attention is a blue alarm. This alarm, which is also known as an alarm indicating signal (AIS), is generated by a higher-order system (HOS), such as a T1C system operating at 3.152 Mbps.

Figure 4.6 illustrates the generation of blue alarms when a failure between higher-order systems occurs. This alarm is generated by each HOS system to channel banks or multiplexers and, in effect, tells each lower-ordered system (DTE) that the problem is between higher ordered systems. Thus, this alarm is primarily used by communications carriers and avoids the dispatching of maintenance personnel to lower-order system facilities. Normally, a blue alarm is generated after 150 ms of loss of an incoming signal. This alarm is produced by transmitting a continuous ones pattern across all 24 channels.

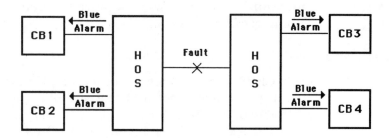

Figure 4.6 Blue alarm generation: HOS = higher order system, CB = channel bank

The reader should note that the previously discussed T1 alarms are not implemented on all T1 circuits. As an example, a point-to-point T1 circuit that is directly routed between two locations and is not switched nor multiplexed by a communications carrier will not generate the previously discussed alarms. However, if the end-user installs equipment that is capable of recognizing alarm conditions and generating those alarms, the T1 line will then pass those alarms generated by their equipment.

In addition to the previously described alarm conditions, there are several error conditions that can be detected by appropriate equipment connected to a T1 circuit. These error conditions include loss of carrier, bipolar violations, and Fs and Ft bit errors.

A loss of carrier condition is defined when receive data is zero for 31 consecutive bits. A bipolar violation is a failure to meet the AMI T1 line code in which marks (1s) are transmitted alternately as positive or negative pulses, while zeros are transmitted as zero volts. An Fs bit error indicates that a signaling framing bit is in error, while an Ft bit error indicates that a terminal framing bit is in error.

4.1.2 CEPT PCM-30 format

CEPT PCM-30 is a PCM format used for time division multiplexing of 30 voice or data circuits onto a single twisted pair cable using digital repeaters. Each voice circuit is sampled at 8 KHz using an 8-bit A-law companding analog-to-digital converter and multiplexed with 29 other sampled channels plus one alignment and one signaling channel, resulting in 32 multiplexed channels.

The standard CEPT frame is 32 channels × 8 bits/channel or 256 bits. With 8000 samples per second, the CEPT data rate becomes 8000 × 256, or 2.048 Mbps. Note that under this format there are no framing bits added to a frame as done under the North American T1 format. This is because the framing bits are carried within specific time slots.

Frame composition

Each CEPT PCM-30 frame consists of 32 time slots to include 30 voice, one alignment and one signaling, with each time slot represented by 8 bits. Since each PCM channel is sampled 8000 times per second, the standard CEPT-30 data rate is 32 × 8 × 8000, or 2.048 Mbps.

Alignment signal

An alignment signal (0011011) is transmitted in bit positions 2 to 8 of time slot 0 in alternating frames. This signal is used to enable each channel to be distinguished at the receiver. Bit position 1 in time slot 0 carries the international bit, while frames not containing the frame alignment signal are used to carry national and international signaling and alarm indication for loss of frame alignment. Figure 4.7 illustrates the composition of the CEPT-30 frame and multiframe, where the multiframe consists of 16 frames, numbered from frame 0 to frame 15.

To avoid imitation of the frame alignment signal, alternating frames fix bit 2 to a 1 in time slot 0 which is the reason why a 1 is entered into that bit position for odd time slot 0 frames.

Signaling data

Time slot 16 in each CEPT-30 frame is used to transmit such signaling data as on-hook and off-hook conditions, dialing digits and call progress. This is indicated by the characters ABCD for frames 1 to 15 in time slot 16 illustrated in Figure 4.7. Since a common channel is dedicated for the signaling data of all-voice circuits, this method of signaling is referred to as common channel signaling and enables each time slot to operate at 64 kbps. In comparison, T1 uses bit robbing to pass signaling information which reduces the effective data rate of time slots to 56 kbps. To enable each frame in a multiframe to be distinguished at the receiver for the recovery of ABCD signaling CEPT-30 uses a multiframe alignment signal. This signal, denoted by the symbol MAS in Figure 4.7, is transmitted in bit positions 1 through 4 of time slot 0 of frame 0. The use of each of the 32 CEPT PCM-30 time slots and their numbering is summarized in Table 4.4.

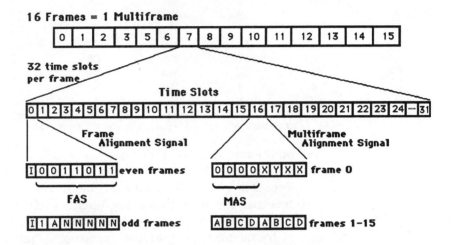

Figure 4.7 CEPT PCM-30 frame and multiframe composition: I = international bit, N = national bit, A = alarm indication signal, FAS = frame alignment signal, ABCD = ABCD signaling bits, X = extra bit for signaling, Y = loss of multiframe alignment, MAS = multiframe alignment signal

CEPT alarms and error conditions

The principal alarms defined by the CEPT PCM-30 format include a red alarm which is produced by a receiver to indicate that it has lost frame alignment and a yellow alarm which is returned to the transmitting terminal to report a loss of frame alignment at the receiving terminal. Both of these alarms function the same, as previously described in the section covering the T1 carrier.

Both red and yellow alarms are generated through the use of the alarm indication signal bit (bit 3) in time slot 0 (TS0) of odd frames. A red alarm is generated by setting bit 3 = 1 in TS0 of non-frame alignment frames. A receiver then indicates the reception of a red alarm by generating a yellow alarm by setting bit 3 = 1 in TS0 of non-frame alignment frames.

Table 4.4 CEPT PCM-30 time slot and channel numbering.

Time slot	Channel	Data use	Time slot	Channel	Data use
0	FAS	No	16	MAS	No
1	1	Yes	17	16	Yes
2	2	Yes	18	17	Yes
3	3	Yes	19	18	Yes
4	4	Yes	20	19	Yes
5	5	Yes	21	20	Yes
6	6	Yes	22	21	Yes
7	7	Yes	23	22	Yes
8	8	Yes	24	23	Yes
9	9	Yes	25	24	Yes
10	10	Yes	26	25	Yes
11	11	Yes	27	26	Yes
12	12	Yes	28	27	Yes
13	13	Yes	29	28	Yes
14	14	Yes	30	29	Yes
15	15	Yes	31	30	Yes

FAS = Frame alignment signal, MAS = multiframe alignment signal.

Two additional alarms generated by CEPT PCM-30 include a multiframe red alarm and a multiframe yellow alarm. The multiframe red alarm is produced by a receiver to indicate that it has lost the multiframe alignment, while the multiframe yellow alarm is returned to the transmitting terminal to report a loss of frame alignment at the receiving terminal. A receiver loses multiframe alignment due to either the occurrence of two consecutive errors in the multiframe alignment signal or, if time slot 16 contains all zeros for at least one multiframe, causing the red alarm to go high, which is coupled to a yellow alarm generator (bit 6 = 1 in time slot 16, frame 0).

CEPT CRC option

For enhanced error monitoring capability, CEPT PCM-30 includes a CRC-4 option. Under this option, a group of eight frames known as a submultiframe is treated as a long binary number. This number is multiplied by X^4 (10000) and divided by $X^4 + X + 1$ (10011). The 4-bit remainder is transmitted in bit position 1 (I bit in Figure 4.7) in time slot 0 in even frames which contain the frame alignment signal. After the receiver computes its own CRC-4 check, it uses bit position 1 in time slot 0 of frames 13 and 15 for CRC error performance reporting. Table 4.5 summarizes how these bits are used for CRC error performance reporting purposes.

Table 4.5   CEPT PCM-30 CRC error performance reporting.

Bit 1 Frame 13	Bit 1 Frame 15	
1	1	CRC for SMF I, II error free
1	0	SMF II in error, SMF I error free
0	1	SMF II error free, SMF I in error
0	0	Both SMF I and II in error

The principal CEPT PCM-30 error conditions include the occurrence of bipolar violations, frame alignment errors, and multiframe alignment errors. A bipolar violation is a failure to meet the AMI CEPT PCM-30 line code, where marks alternate as positive and negative pulses and spaces are represented by a zero voltage. A frame alignment error is a failure to synchronize on the frame alignment signal (0011011) contained in time slot 0 of alternating frames, while a multiframe alignment error is a failure to synchronize on the multiframe pattern (0000) contained in bits 1–4 of time slot 16 of frame 0.

4.2 T-CARRIER SIGNAL CHARACTERISTICS

As previously discussed in Chapter 1, there are several advantages to transmitting T-carrier signals in a bipolar alternate mark inversion (AMI) format, including the absence of a dc component to the signal and the ability to obtain clock recovery from the signal in an all-ones condition. Unfortunately, the disadvantage associated with this signaling method is that a sequence of spaces is encoded as a period of zero voltage or no signal, and repeaters on a span line cannot recover clocking without a signal occurring every so often.

For repeaters to properly recover clocking, a certain number of binary ones must be contained within a transmitted signal. Since North American and European approaches to this problem differ, we will examine the encoding methods used to insure that an appropriate signal contains a minimum number of binary ones for each location.

4.2.1 North America

In North America, AT&T publication 62411 sets the ones density requirement to be 'n' ones in each window of $8 \times (n + 1)$ bits, where n varies from 1 to 23. This means that a T1 carrier cannot have more than 15 consecutive zeros ($n = 1$) and there must be approximately three ones in every 24 consecutive bits ($n = 2$ to 23). Several methods are currently used to provide this minimum ones density, including binary 7 zero code suppression, binary 8 zero substitution, and zero byte time slot interchange.

Binary 7 zero code suppression

Under the binary 7 zero code suppression method a binary one is substituted in bit position 7 of each time slot if all eight positions are zeros. An example of this method of insuring a minimum ones density is illustrated in Figure 4.8A. Although it might appear wiser to select the least significant bit for inversion, this cannot be done since the setting of a frame bit to zero, when bit positions 2 through 8 in the previous time slot were set to zero, would result in a string of 16 consecutive zeros if the bits in the time slot following the frame bit were zero and bit position 8 was used for substitution. Figure 4.8B illustrates this worst-case scenario which explains why bit position 7 in each time slot is used for bit value inversion to insure a minimum ones density.

Effect upon bit robbing

If a data channel contains all 0s, the data can be corrupted due to B7 zero suppression. Due to this, a data channel normally is restricted to seven usable data bits, with one bit in the data channel set to a 1. This prevents the data channel from being corrupted, but also limits its data rates to 56 kbps.

When one bit is set to a 1 on a DS0 channel, the channel is known as a non-clear channel. The 56 kbps on a non-clear channel is also known as a DS-A channel.

A T1 clear channel is one in which all 64 kbps in each DS0 are usable. On private microwave systems, B7 zero code suppression is normally not required, permitting clear channel capability.

Binary 8 zero substitution

The binary 8 zero substitution (B8ZS) technique was developed by Bell Laboratories and is now sanctioned by the CCITT for use in North America. This method of insuring a minimum ones density was placed into operation during the mid-1980s and offers

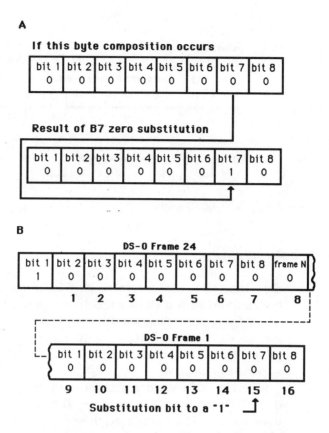

Figure 4.8 B7 zero code suppression: A B7 zero code suppression example; B worst-case scenario

a significant improvement over binary 7 zero code suppression, as it both maintains a minimum ones density and also provides a clear channel capability, permitting each DS0 channel to carry data at 64 kbps. Under B8ZS coding, each eight consecutive 0s in a byte are removed and replaced by a B8ZS code. If the pulse preceding an all-zero byte is positive, the inserted code is 000+−0−+. If the pulse preceding an all-zero byte is negative, the inserted code is 000−+0+−. Figure 4.9 illustrates the use of B8ZS coding in which an all-zeros byte is replaced by one of two binary codes, with the actual code used based upon whether the pulse preceding the all-zeros byte was positive or negative.

Both examples result in bipolar violations occurring in the fourth and seventh bit positions. Both carrier and customer equipment must recognize these codes as legitimate signals, and not as bipolar violations or errors, for B8ZS to work to enable a receiver to recognize the code and restore the original eight zeros

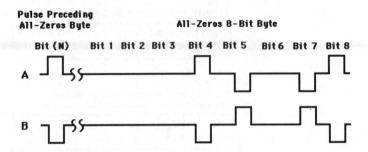

Figure 4.9 B8ZS coding

Zero byte time slot interchange

Zero byte time slot interchange (ZBTSI) is a method of encoding information into a PCM word to prevent an excessive number of zeros from occurring on the transmission line. Unlike the two previously discussed zero suppression methods that can be used in any framing method, ZBTSI is available only with an extended superframe format.

When a PCM word containing all zeros is found, a CSU with a ZBTSI encoder will replace the zero byte with addressing information, while overhead information is transmitted by using 2 kbps of the 4 kbps ESF data link. The encoder operates on 96 PCM words at a time, representing four ESF frames. The data link bits from frames 1, 5, 9, 13, 17, and 21 are used to provide encoding flags for the 96 words following each flag bit and are referred to as z bits. Figure 4.10 illustrates the relationship of the ZBTSI flag bits with respect to the ESF frames.

Currently, ZBTSI is supported by several Bell Operating Companies; however, it appears that binary 8 zero substitution has a wider adoption by communications carriers and will probably gain additional acceptance in the future over the other two zero suppression methods.

4.2.2 Europe

In Europe the high density bipolar 3-zero maximum (HDB3) coding is used by CEPT PCM-30 to obtain a minimum ones density for clock recovery from received data. Under HDB3, the data stream to be transmitted is monitored for any group of four consecutive

ESF Frame Number	ESF Bit Number	Framing Bits Assignments		
		F3	Data Link DL	ZBTSI
CRC				
1	0	-	-	z	-
2	193	-	-	-	CB1
3	386	-	d	-	-
4	579	0	-	-	-
5	772	-	-	z	-
6	965	-	-	-	CB2
7	1158	-	d	-	-
8	1351	0	-	-	-
9	1544	-	-	z	-
10	1737	-	-	-	CB3
11	1930	-	d	-	-
12	2123	1	-	-	-
13	2316	-	-	z	-
14	2509	-	-	-	CB4
15	2702	-	d	-	-
16	2895	0	-	-	-
17	3088	-	-	z	-
18	3281	-	-	-	CB5
19	3474	-	d	-	-
20	3667	1	-	-	-
21	3860	-	-	z	-
22	4053	-	-	-	CB6
23	4246	-	d	-	-
24	4439	1	-	-	-

Figure 4.10 ESF framing bit assignments with ZBTSI: Fe = framing pattern sequence, DL = 4 kbps data link channel, ZBTSI = ZBTSI encoding flag bits (Z bits), CRC = cyclic redundancy check field

zeros. A four-zero group is then replaced with an HDB3 code. Two different HDB3 codes are used to insure that the bipolar violation pulses from adjacent four-zero groups are of opposite polarity as indicated in Figure 4.11. The selection of the HDB3 code is based upon whether there was an odd or even number of ones since the last bipolar violation (BV) occurred. If an odd number of ones

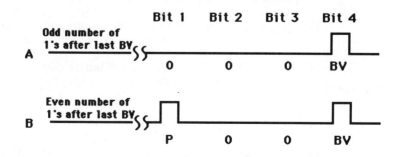

Figure 4.11 HDB3 coding: P = polarity bit, BV = bipolar violation

occurred since the previous bipolar violation, the coding method in Figure 4.11A, is used to replace a sequence of four zeros. If an even number of ones occurred since the previous bipolar violation, the coding method in Figure 4.11B is used to replace a sequence of four zeros.

REVIEW QUESTIONS

1 Describe the composition of a DS1 frame, including the placement of each DS0 PCM word and the framing bit.

2 What is the actual data rate available for use on a DS1 signal?

3 How many bits are included in a superframe?

4 Discuss two advantages of the use of the extended superframe format over D4 framing.

5 How many bits are in an extended superframe?

6 Discuss the use of each of the three types of frame bits in the extended superframe.

7 What does a mismatch between the locally generated check bit sequence and the received check bit sequence under the extended superframe format indicate?

8 Discuss the relationship between a yellow and a red alarm.

9 Where does a blue alarm originate and what does it indicate?

10 How many bits are in a CEPT frame? Where are the framing bits in a CEPT frame?

11 How many bits are contained in the CEPT multiframe?

12 Why can a CEPT DS0 carry data at 64 kbps, while a T1 DS0 is normally limited to a 56 kbps data rate?

13 Why was the seventh bit in a PCM word selected for inversion under binary 7 zero code suppression instead of the least significant bit?

14 What is the common advantage of binary 8 zero substitution and zero byte time slot interchange zero suppression methods over binary 7 zero code suppression?

5

T-CARRIER MULTIPLEXERS

The key to the effective utilization of T-carrier transmission facilities is multiplexers. As previously noted in Chapter 2, telephone company channel banks contain time division multiplexers. Two other types of multiplexing equipment that can be effectively used with T-carrier facilities are terminal or end-unit multiplexers and nodal switches. In this chapter we will examine the operation and utilization of each category of T-carrier multiplexing equipment and how they can be used to support a variety of end-user transmission requirements.

5.1 CHANNEL BANKS

The channel bank represents the earliest type of multiplexing equipment used to interface a T-carrier facility. Channel banks are used by telephone companies to multiplex either 24 or 32 digitized voice channels onto a T-carrier.

The channel bank was developed by Bell Laboratories and has been used since the 1960s in telephone company central offices with a digital access cross connect (DACS) system to switch digitized voice calls.

5.1.1 Evolution

In the United States, the first set of digital channel banks was known as D1 systems, where the D stands for digital which was a significant difference in multiplexing in comparison to the prior method of using frequency division multiplexing. The D1 channel bank used a seven-bit PCM algorithm for digitizing voice and the eighth bit of each eight-bit channel time slot was used for signaling. The D1 channel bank multiplexed the eight bits of each of the 24 channels into a multiplexing frame and added a framing bit to maintain synchronization between channel banks. The framing bit was used to generate a repeating '1010' pattern which is referred to as D1 framing.

The next evolution in the development of carrier multiplexing equipment was the D2 channel bank. The D2 channel bank extended the digitization of voice into an eight-bit PCM word and introduced the bit robbing technique discussed in Chapter 4 to obtain a signaling capability to pass such channel signaling information as on-hook, off-hook, and dialed digits. To maintain better synchronization, the D2 channel bank employed a pseudo-random time slot sequencing of digitized voice time slots into the 24 channels in the frame. Figure 5.1 illustrates the channel number sequence assignment used by D2 channel banks.

Time
Slot   1   2 3 4   5   6   7   8   9 10 11 12 13 14 15 16 17 18 19 20 21 22 23 24

| 12 | 13 | 1 | 17 | 5 | 21 | 9 | 15 | 3 | 19 | 7 | 23 | 11 | 14 | 2 | 18 | 6 | 22 | 10 | 16 | 4 | 20 | 8 | 24 |

Channel
Assignment

Figure 5.1 D2 channel number sequence assignment. D2 channel banks employ a pseudo-random assignment of the 24 DS0 channels into the 24 time slot positions of the frame

The D3 channel bank is essentially the same as the D2 channel bank but changed the time slot sequencing in the frame to correspond to the channel numbering sequence. Included in this channel bank was circuitry that provided a minimum ones density as discussed in Chapter 4.

The D4 channel bank was placed into service during 1977 and is still used today. This channel bank multiplexes two sets of 24 channels onto two T1 circuits or it can be used to interleave two T1 data streams onto a T1C circuit operating at 3.152 Mbps. Due to the use of bit robbing, data rates are restricted to a maximum of 56 kbps on each DS0 channel of a D4 channel bank.

Each D4 channel bank requires the installation of specific circuit cards to provide an interface to a specific type of voice or data circuit. As an example, a 56 kbps office channel unit (OCU) and a 56 kbps DSU are required to interface a 56 kbps data circuit onto a

DS0 channel, while a four-wire E&M (ear and mouth) circuit card must be used to interface the channel bank to a four-wire E&M circuit.

The D4 channel bank is divided into two groups called di-groups, with each group containing 24 channels as illustrated in Figure 5.2. The function of the common equipment for each di-group is listed in Table 5.1.

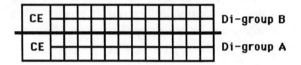

Figure 5.2 D4 channel bank: CE = Common equipment

Table 5.1 D4 channel bank functional units.

Unit	Operational function
Transmit unit	Performs the function of directing channel sampling encoding and insertion of framing pulses for the di-group
Receive unit	Performs the function of decoding the PCM signal for a di-group, demultiplexing the channel information and extracting the timing, framing and signaling information
Trunk processing unit	Automatically disconnects customers from a circuit during a carrier failure, making the circuits busy so customers cannot seize them, as well as stops all charges on the calls
Power distribution unit	Distributes power to the channel bank
Power control unit	Functions as a dc-to-dc converter to supply correct voltage to the channel bank
Office interface unit	Provides or derives one of three types of clocking to include:
Local timing	Timing is developed within the channel bank and is independent of any other source
Looped timing	Timing is derived from the received signal and is then used for the transmitted signal
External timing	Timing is obtained from an external device

5.1.2 Channel banks versus T-carrier multiplexers

The major differences between channel banks and T-carrier multiplexers are in the areas of voice interfaces, diagnostic capability; and the ability to perform automatic rerouting of data. Although channel banks support a large variety of voice interfaces, they have limited diagnostics and cannot be used for rerouting. In comparison, T-carrier multiplexers may have a limited voice interface capability; however, they usually have superior diagnostics and many provide the capability to automatically reroute data.

5.1.3 Special carrier multiplexing facilities

AT&T currently offers two special types of multiplexing service marketed under the names M24 and M44. Since equivalent services are offered by other communications carriers, we will review their functionality in this section.

M24 multiplexing

M24 multiplexing is a service provided under Accunet T1.5 which allows a T1 line to be broken out into 24 individual lines. The resulting individual lines can access the PSTN, DDS service, or leased lines as illustrated in Figure 5.3. Note that under AT&T's M24 service, a digital access and cross connect system (DACS) is used to route specific DS0 channels to a variety of other carrier facilities as indicated in the previously referenced illustration.

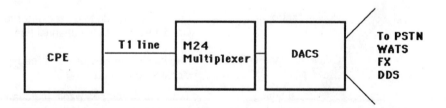

Figure 5.3 M24 multiplexing. M24 multiplexing permits a customer to access a wide assortment of carrier facilities via a common T1 line: CPE = customer premises equipment, DACS = digital access cross connect system, PSTN = public switched telephone network, FX = foreign exchange, DDS = Dataphone Digital Service

M44 multiplexing

M44 multiplexing is a service which compresses voice signals from two T1 lines onto one T1 line and expands compressed voice signals back into two T1 lines.

The key to M44 multiplexing is the use of adaptive differential pulse code modulation (ADPCM) at the carrier's central office where M44 multiplexing is performed. At that location 22 DS0 channels on each of two T1 lines are first demultiplexed by a DACS. The resulting 64 kbps channels are passed to an M44 multiplexer which encodes each channel according to the ADPCM algorithm and returns the data to the DACS which then bundles 44 DS0 channels onto a T1 circuit. Figure 5.4 illustrates the M44 multiplexing process.

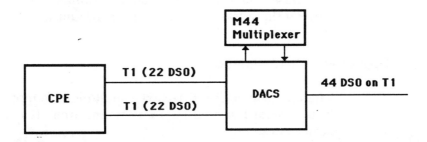

Figure 5.4 M44 multiplexing. M44 multiplexing combines 44 DS0 channels onto one T1 circuit through the use of adaptive differential pulse code modulation

M44 service can be used between central offices or between a central office and a customer's premises. Although this service offers a significant economic improvement over the use of PCM encoding, it is limited to compressing only voice, requiring data to be passed through uncompressed.

5.2 T-CARRIER MULTIPLEXERS

A T-carrier multiplexer is a device that can be used to integrate voice, data and video onto a T1 link.

T1 facilities were first offered to the public as a tariffed service by AT&T in 1983. The availability of AT&T's Accunet T1.5 service was a driving force in a literal explosion of new and established companies introducing products for use with this service.

T-carrier multiplexers were originally point-to-point devices with each multiplexer considered to be a terminal or end-unit device that neither provided networking capability nor the ability to dynamically assign bandwidth utilization. Commensurate with the growth in T-carrier networking has been a corresponding increase in the features and capabilities of T-carrier multiplexers. Many T-carrier multiplexers now include multinodal support capability, the ability to perform many types of voice digitization through the addition of voice digitization modules as well as the ability to dynamically assign voice, data, and video to the T-carrier bandwidth.

Since the primary difference between a T-carrier terminal or end-unit multiplexer and a T-carrier nodal switch is in the areas of multi-trunk support capability and the automatic rerouting of data, we will first focus our attention upon their common operational characteristics and features in this section. Using this information as a base will then enable us to examine the key differences between these two types of T-carrier multiplexers.

5.2.1 Operational characteristics

A minimally configured T-carrier multiplexer supports from 24 or 30 to several hundred DS0 inputs and from 1 to 12 or more T-carrier lines.

Some T-carrier multiplexers digitize voice directly through the addition of optional voice digitizer modules. Other T-carrier multiplexers require digitized voice to be routed as input to the multiplexer. Figure 5.5 illustrates a typical T-carrier multiplexer application where the device is used to combine a variety of digitized voice, data, and video inputs onto a T-carrier facility operating at a 1.544 Mbps (North American) or 2.048 Mbps (European) data rate.

In the example illustrated in Figure 5.5, it was assumed that the digitized video conferencing required the use of 11 DS0 channels. Hence, the operating rate required to support full motion video through the multiplexer becomes 64 kbps times 11, or 704 kbps. The ten lines routed from the PBX are assumed to be analog, resulting in a requirement for voice digitizer modules to be installed in the T-carrier multiplexer. In this example, PCM digitization modules were used, resulting in each analog channel digitized at 64 kbps onto one DS0 channel. Thus, the ten voice channels are shown as collectively occupying 640 kbps of the T-carrier bandwidth.

In the lower portion of Figure 5.5, twelve 4.8 kbps data sources are first multiplexed by a conventional TDM, resulting in the composite 57.6 kbps data stream boosted to 64 kbps by the use of pad bits to enable its support by one T-carrier channel. The use of this type of multiplexer to pre-multiplex low data rate asynchronous or synchronous data sources will depend upon the functionality of the T-carrier multiplexer and its ability to multiplex subrate digital data streams. Some T-carrier multiplexers are limited to multiplexing synchronous data only. Other T-carrier multiplexers may support a variety of asynchronous and synchronous data rates through the use of different channel cards. Table 5.2 lists some of the more common data rates supported by T-carrier multiplexers, including the operating rate of many T-carrier voice digitization modules.

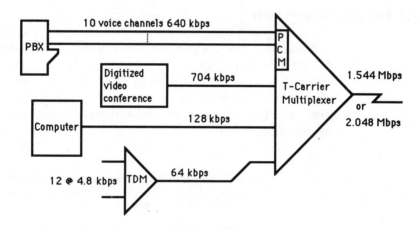

Figure 5.5 Typical T1 multiplexer application

Table 5.2 Typical T1 multiplexer channel rates.

Type	Data rates (bps)
Asynchronous	110; 300; 600; 1200; 1800; 2400; 3600; 4800; 7200; 9600; 19 200
Synchronous	2400; 4800; 7200; 9600; 14 400; 16 000; 19 200; 32 000; 38 400; 40 800; 48 000; 50 000; 56 000; 64 000; 112 000; 115 200; 128 000; 230 400; 256 000; 460 800
Voice	16 000; 32 000; 48 000; 64 000

5.2.2 Multiplexing efficiency

In addition to examining the type of voice, data, and video support, it is also important to determine how efficiently a T-carrier multiplexer utilizes each DS0 channel. Some T-carrier multiplexers can only place one asynchronous or synchronous data source onto a DS0 channel regardless of its data rate; whereas, other multiplexers may make much more efficient utilization of DS0 channels. In our examination of T-carrier multiplexer features, which follows this section, we will investigate the advantages of a subrate multiplexing capability.

5.2.3 Features to consider

In Table 5.3 the reader will find a list of the major features of T-carrier multiplexers that warrant consideration during an acquisition process. While all of the listed features are important to consider, they may not be relevant to certain situations based upon the immediate and long-term requirements of a specific organization.

Table 5.3   T-carrier multiplexer features to consider.

Bandwidth utilization method

Bandwidth allocation method

Voice interface support

Voice digitization support

Internodal trunk support

Subrate channel utilization

Digital access cross connect capability

Gateway operation support

Alternate routing and route generation

Redundancy

Maximum number of hops and nodes supported

Diagnostics

Configuration rules

Bandwidth utilization

Inefficient T-carrier multiplexers assign data to the T-carrier facility using 64 kbps DS0 channels for each data source as illustrated in Figure 5.6A. In this example the assignment of input data sources is fixed to predefined channels, resulting in the inability of the multiplexer to take advantage of the inactivity of different data sources.

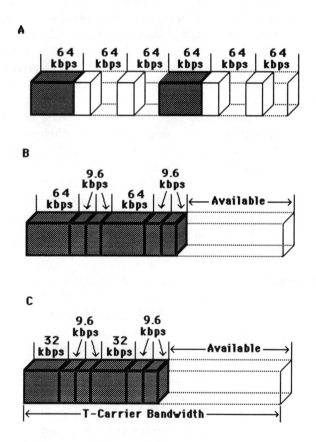

Figure 5.6 Demand assigned bandwidth: A conventional bandwidth allocation with PCM digitization; B demand assigned bandwidth with PCM digitization; C demand assigned bandwidth with ADPCM digitization

More efficient T-carrier multiplexers employ a variety of demand assigned bandwidth techniques to make more efficient use of the composite T-carrier bandwidth. This is illustrated in Figure 5.6B in which a basic demand assignment feature of a T-carrier multiplexer dynamically assigns bandwidth based upon the activity of the data sources. In this example it was assumed that several 9.6 kbps data sources became active along with two PCM digitized voice conversations and were dynamically assigned to the T-carrier bandwidth in their order of activation. Note that this method of bandwidth assignment normally results in an increase in available bandwidth since the probability of all inputs becoming active at one time is usually very low. In addition, having the capability to allocate bandwidth based upon the data rate of the data sources and not by a DS0 channel basis allows the 9.6 kbps data sources to occupy significantly less bandwidth. Thus, demand assignment with dynamic bandwidth allocation results in a considerable improvement in the use of a T-carrier's data transmission capacity in comparison to a conventional bandwidth allocation process.

Figure 5.6C illustrates the effect upon bandwidth allocation based upon the use of a more efficient voice digitization module in T-carrier multiplexers. In this example it was assumed that ADPCM voice digitization modules were used in the T-carrier multiplexer instead of PCM voice digitization modules. The use of ADPCM reduces the bandwidth required for carrying each voice conversation to 32 kbps, further increasing the available bandwidth of the T-carrier to support other data sources.

Bandwidth allocation

Most T-carrier multiplexers use time division multiplexing schemes to allocate bandwidth to each voice and data channel as well as portions of DS0 channels. Techniques used for bandwidth allocation can include the demand assignment of bandwidth previously illustrated in Figure 5.6B and C, as well as non-contiguous resource allocation and the packetization of voice, data, and video.

Figure 5.7 illustrates the advantage of non-contiguous resource allocation over the conventional method of allocating DS0 channels. In Figure 5.7A, a section of T-carrier bandwidth supporting three voice calls is shown. Here, each call is placed in a contiguous portion of the T-carrier bandwidth. In Figure 5.7B it was assumed that call B was completed and its bandwidth became available for use. Now suppose a data source (D) became active that required more bandwidth than that freed by the completion of call B. Under the non-contiguous resource allocation method the bandwidth required to accommodate the data source could be split into two or more non-contiguous sections of the T-carrier bandwidth as illustrated in Figure 5.7C.

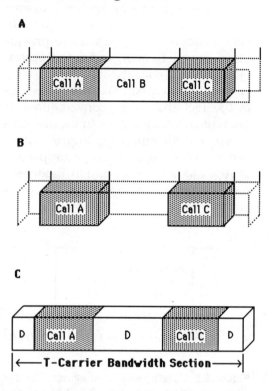

Figure 5.7 Bandwidth allocation methods. Non-contiguous resource allocation of bandwidth enables input to the T-carrier multiplexer to be split into portions of the available bandwidth: A T-carrier bandwidth section supporting three calls; B call B is completed and its bandwidth becomes available; C Data source D multiplexed into non-contiguous sections of bandwidth

The third method of bandwidth assignment was also pioneered by Stratacom with that firm's introduction of a T-carrier multiplexer that packetizes both voice and data. The Stratacom multiplexer uses 'fast packet' technology where the term fast packet refers to the fact that information is transmitted across the network in packet format instead of a time division multiplexed format. Although the external interface to voice and data is the same as a conventional T-carrier multiplexer, the internal operation of the Stratacom multiplexer is considerably different from other devices.

Another interesting method of bandwidth allocation involves the packetization of voice and data sources onto T-carrier facilities. This technique was also pioneered by Stratacom.

The Stratacom fast packet multiplexer generates packets only when data sources are active, using a packet length of 193 bits which corresponds to the North American T1 frame length. Figure 5.8 illustrates the Stratacom frame format. Although 20 bits, in effect, function as overhead to provide a destination address (16 bits), priority (2 bits), and error correction to the header by the use of a hamming code (6 bits), the efficiency of packetized multiplexing can be considerable. This is because the technique takes advantage of the fact that voice conversations have periods of silence and are typically half-duplex in nature. This enables packet technology to provide an efficiency improvement of approximately 2:1 over conventional time division multiplexing of voice. With the addition of ADPCM voice digitization modules, the Stratacom fast packet multiplexer can support up to 96 voice conversations on a T1 circuit.

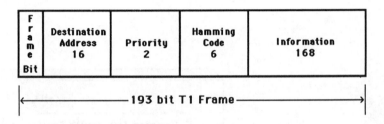

Figure 5.8 Stratacom packet format. The Stratacom T1 multiplexer packetizes voice and data sources into 193 bit frames containing 168 information bits

One of the problems associated with the use of packet technology to transport digitized voice is the fact that you cannot delay voice. Thus, unlike data packets that can be retransmitted if an error is detected, packetized voice cannot tolerate the delay of retransmission. In addition to not being able to retransmit voice, one must also consider the effect of a large number of voice channels becoming active. When too many channels become active, the total bit rate of the digitized input channels can exceed the output bit rate of the T1 circuit. To avoid too much delay to specific channels, some channels will be skipped since a listener can tolerate a 125-millisecond delay. Another problem associated with packetized voice is the delay that can occur as the packets are routed from node to node in a complex network. To overcome this problem, Stratacom incorporates a priority field in its packet which enables certain packets to be processed and routed before other types of packets.

Voice interface support

Since most T1 multiplexer applications include the concentration of voice signals, the type of voice interfaces supported for two-wire and four-wire applications is an important multiplexer feature to consider. Prior to examining the types of voice interfaces supported by T1 multiplexers, let us first review some of the more common types of voice signaling methods since it is the signaling method that is actually supported by a particular interface.

Two of the most common types of telephone signaling include loop signaling and E&M signaling. Loop signaling is a signaling method employed on two-wire circuits between a telephone and a PBX or between a telephone and a central office. E&M signaling is a signaling method employed on both two-wire and four-wire circuits routed between telephone company switches.

In loop signaling, the raising of the telephone handset results in the activation of a relay at the PBX or central office, causing current to flow in a circuit formed between the telephone set and the PBX or central office. The raising of the handset, referred to as an off-hook condition, results in the PBX or central office returning a dial tone to the telephone set. As the subscriber dials the telephone number of the called party, the dialed digits are received at a telephone company central office which then signals the called party by sending signaling information through the telephone company network. Once the call is completed the placement of the handset back onto the telephone set, a condition known as on-hook, causes the relay to be deactivated and the circuit previously formed to open.

A second type of telephone off-hook signaling that flows in a loop is ground start signaling. This method of signaling is also used on two-wire circuits between a telephone set and a PBX or central office. Unlike loop start signaling, in which loop seizure is detected at the PBX or central office, ground start allows the detection of loop seizure to occur from either end of the line.

E&M signaling

E&M signaling is used on both two-wire and four-wire circuits connecting telephone company switches. Here the M lead is used to send ground or battery signals to the signaling circuits at a telephone company switch, while the E lead is used to receive an open or ground from the signaling circuit. In E&M signaling the local end asserts the M lead to seize control of the circuit. The remote end receives the signal on the E lead and toggles its M lead as a signal for the local end to proceed. The local end then sends the address by toggling its M lead, in effect, placing dialing pulses on that lead which is used by the remote end to effect the desired connection. Once a call is completed, either party will drop its M lead, resulting in the other side responding by dropping its M lead. Currently, there are three types of E&M signaling: types I, II, and III. The difference between E&M signaling types relates to the method by which an on-hook condition is established—ground or open.

Table 5.4 lists some of the more common types of voice interface cards supported by many T-carrier multiplexer vendors. The two-wire and four-wire transmission only interfaces are designed to support permanent two-wire and four-wire connections between two points that do not require the passing of signaling information. Both types of interfaces are normally used to support a data modem connection through a T-carrier multiplexer.

Table 5.4 Typical T-carrier voice interface modules.

Two-wire transmission only
Two-wire E&M
Two-wire foreign exchange
Four-wire transmission only
Four-wire E&M

The two-wire and four-wire E&M interfaces usually support the connection of PBXs and telephone company equipment to a T-carrier multiplexer. As previously mentioned, there are three types of E&M signaling, with each type applicable to both two-wire and four-wire operations.

The two-wire foreign exchange office interface is designed to support the attachment of a T-carrier multiplexer to a PBX or central office switching equipment that provides an open or closed foreign exchange termination point.

Voice digitization support

In addition to the method of bandwidth assignment and allocation, a third major feature affecting the efficiency of a T-carrier multiplexer is the type of voice digitization modules the device supports. Although most T-carrier multiplexers support the use of PCM and ADPCM, some vendors also support the use of adapter cards that contain proprietary voice digitization modules. One example of this is adaptive speech interpolation which changes the digitization rate of selected voice channels from 32 kbps to 24 kbps as available bandwidth becomes saturated. Although some proprietary techniques may offer advantages in both the fidelity of a reconstructed voice signal as well as in the bandwidth required to carry the signal, their use restricts an organization to one vendor's product.

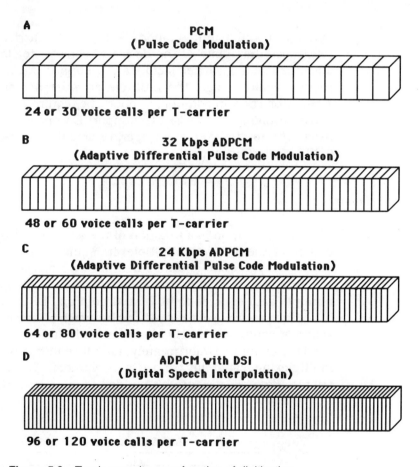

A

PCM
(Pulse Code Modulation)

24 or 30 voice calls per T-carrier

B

32 Kbps ADPCM
(Adaptive Differential Pulse Code Modulation)

48 or 60 voice calls per T-carrier

C

24 Kbps ADPCM
(Adaptive Differential Pulse Code Modulation)

64 or 80 voice calls per T-carrier

D

ADPCM with DSI
(Digital Speech Interpolation)

96 or 120 voice calls per T-carrier

Figure 5.9 Trunk capacity as a function of digitization

Figure 5.9 illustrates the effect of the use of several types of voice digitization modules upon the capacity of a T-carrier. If standard PCM modules are used, the T-carrier becomes capable of supporting either 24 or 30 voice calls depending upon whether a North American or European T-carrier facility is used. When 32 kbps ADPCM modules are used to digitize voice, the voice carrying capacity of the T-carrier is doubled as shown in Figure 5.9B. Figure 5.9C, which illustrates the use of 24 kbps ADPCM, shows the voice carrying capacity of a T-carrier tripling, while Figure 5.9D shows how the voice carrying capacity of a T-carrier can be quadrupled through the use of ADPCM and DSI.

Internodal trunk support

The internodal trunk support feature of T-carrier multiplexers references the ability of the device to connect to North American and European T-carrier facilities. To support North American T-carrier facility usage, the multiplexer must not only operate at 1.544 Mbps but, in addition, support the required communications carrier framing–D4 or ESF. To support European T-carrier facility usage the multiplexer must operate at 2.048 Mbps and support CEPT PCM-30 framing.

Subrate channel utilization

Channel utilization is a function of the subrate multiplexing capabilities of the T-carrier multiplexer. Many T-carrier multiplexers support asynchronous data rates from 50 bps to 19.2 kbps and synchronous data rates from 2.4 kbps to 19.2 kbps, permitting multiple data sources to be placed onto one DS0 channel. Figure 5.10 illustrates one of the methods by which a T-carrier multiplexer vendor's equipment might multiplex subrate data channels onto one DS0 channel. Unfortunately, not all vendors provide a subrate multiplexing capability in their equipment. When this occurs, subrate data sources are bit padded to operate at 64 kbps which

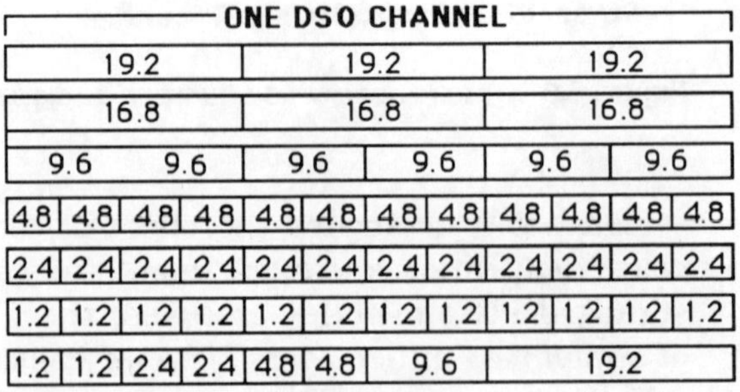

```
┌───────────────── ONE DS0 CHANNEL ─────────────────┐
│      19.2       │      19.2       │      19.2       │
├─────────────────┼─────────────────┼─────────────────┤
│      16.8       │      16.8       │      16.8       │
├──────────┬──────┼──────┬──────────┼──────┬──────────┤
│   9.6    │ 9.6  │ 9.6  │   9.6    │ 9.6  │   9.6    │
├────┬─────┼────┬─┼────┬─┼────┬─────┼───┬──┼───┬──────┤
│4.8 │4.8 │4.8│4.8│4.8│4.8│4.8│4.8│4.8│4.8│4.8│4.8│
├───┬───┬───┬───┬───┬───┬───┬───┬───┬───┬───┬───┤
│2.4│2.4│2.4│2.4│2.4│2.4│2.4│2.4│2.4│2.4│2.4│2.4│
├──┬──┬──┬──┬──┬──┬──┬──┬──┬──┬──┬──┤
│1.2│1.2│1.2│1.2│1.2│1.2│1.2│1.2│1.2│1.2│1.2│1.2│
├───┬───┬───┬───┬───┬───┬──────┬──────────┤
│1.2│1.2│2.4│2.4│4.8│4.8│ 9.6  │   19.2   │
└────────────────────────────────────────┘
```

Figure 5.10 Typical subrate channel utilization

can considerably reduce the ability of the multiplexer to maximize bandwidth utilization. In such situations users can obtain one or more subrate multiplexers to combine several data sources to a 64 kbps data rate; however, this may result in a higher cost than obtaining T-carrier multiplexers that include a built-in subrate multiplexing capability.

Digital access cross connect capability

The ability of a multiplexer to provide digital access cross connect operations can be viewed as the next step up in terms of functionality from point-to-point multiplexer operations. Although communications carrier DACS are limited to switching DS0 channels, many multiplexer vendors include the capability to drop and insert/bypass subrate channels or digitized voice encoded at bit rates under 64 kbps, permitting sophisticated T-carrier networks to be constructed.

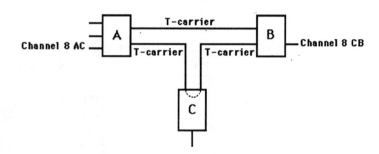

Figure 5.11 DS0 cross connect. Although many multiplexers support the cross connection of DS0 channels, some T-carrier multiplexers also permit the cross connection of subchannels

Figure 5.11 illustrates an example of the use of a digital access cross connect feature used in three T-carrier multiplexers labeled A, B, and C. In this example, channel 8 on multiplexer A is routed to multiplexer C where it is dropped, freeing that DS0 channel for use, as the T-carrier is then routed to multiplexer B's location. Thus, a data source at location C could be inserted into the T-carrier on channel 8, resulting in channel 8 being routed from C to B as shown in Figure 5.11. In this example it was assumed that all other DS0 channels were simply passed through or bypassed multiplexer C and were then routed to multiplexer B. Thus, a digital access and cross connect capability can be used to establish a virtual circuit through an intermediate multiplexer (bypass) without demultiplexing the data, to allow intermediate nodes to add data to the data stream (insert), as well as to permit an intermediate node to act as a terminating node (drop) for other multiplexers.

Gateway operation support

To function as a gateway requires a T-carrier multiplexer to support a minimum of two high-speed circuits. In addition, the T-carrier multiplexer must perform several other operations that must be coordinated with the use of the T-carrier multiplexers connected to the gateway multiplexer.

Three of the main problems associated with connecting European and North American T-carrier circuits through the use of a gateway multiplexer involve compensating for the differences between European and North American T-carriers with respect to the number of DS0 channels each T-carrier supports, the method by which signaling is carried in each channel, and the method by which performance monitoring is accomplished.

Since a European T-carrier contains 30 DS0 channels, while a North American T1 link supports 24, the gateway multiplexer will map 30 DS0 channels to 24, resulting in the loss of six channels. This means that the end-user's organization is limited to the effective use of 24 channels on a European connection via a gateway multiplexer.

For signaling conversion the gateway multiplexer will move AB or ABCD signaling under D4 and ESF frame formats into channel 16 for North American to European conversion. For signaling conversion in the other direction, appropriate bits will be moved from channel 16 to the robbed bit positions used in D4 and ESF framing.

The area of performance monitoring is presently either ignored by gateway multiplexers or handled on an individual link basis to the gateway device. Since European systems use a CRC-4 check while ESF employs a CRC-6 check, no conversion is performed by multiplexers that support performance monitoring, since the results would be meaningless for a link consisting of both North American and European facilities connected through a gateway. Instead, the gateway will provide statistics treating each connection as a separate T-carrier facility. Figure 5.12 illustrates the placement of a gateway T-carrier multiplexer connecting a North American T1 facility to a European CEPT 30 facility.

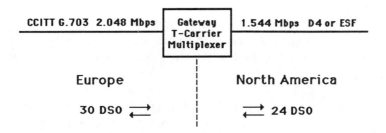

Figure 5.12 Gateway operation. The operation of a gateway results in the dropping of six DS0 channels when a European CEPT 30 facility is converted to a North American T1 link

Alternate routing and route generation

If a T-carrier network consists of three or more multiplexers interconnected by those carrier facilities, both the alternate routing capability and the method of route generation are important features to consider.

Basically, route generation falls into two broad areas: paths initiated by tables constructed by operators and dynamic paths automatically generated and maintained by the multiplexers. To illustrate both alternate routing and route generation, consider the T-carrier network illustrated in Figure 5.13. In this illustration the T-carrier circuit connecting multiplexers A and B has failed. With an alternate routing capability some or all DS0 channels previously carried by circuit AB must be routed from path AC to path CB to multiplexer B.

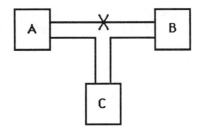

Figure 5.13 Alternate routing. T-carrier multiplexers provide a variety of methods to effect alternate routing to include using predefined tables and by the dynamic examination of current activity at the time of failure

If the multiplexers employ alternate routing based upon predefined tables assigned by operations personnel, DS0 channels previously routed on path AB will be inserted into the T-carrier linking A to C and then routed onto path CB based upon the use of those tables. As the DS0 channels from path AB are inserted into the T-carrier linking A to C, DS0 channels on path AC must be dropped, a process referred to as bumping. If calls were in progress on path A to C and C to B when the failure between A and B occurred, those calls may be dropped, depending upon whether or not the multiplexers employ a priority bumping feature or have the capability to downspeed voice digitized DS0 channels.

Priority bumping refers to the ability to override certain existing DS0 subchannels based upon the priority assigned to DS0 channels that were previously carried on the failed link and the priorities assigned to DS0 active channels on the operational links. Downspeed refers to the capability of multiplexers to shift to a different and more efficient voice digitization algorithm to obtain additional bandwidth with the ability to carry DS0 channels from the failed link on the operational circuits. One example of downspeed would be switching from 32 kbps ADPCM to 24 kbps ADPCM, resulting in freeing up 12 kbps per operational DS0 channel.

When alternate routing and route generation is dynamically performed by multiplexers, those devices examine current DS0 activity and establish alternate routing based upon predefined priorities and the current activity of DS0 channels. If predefined tables are used, the multiplexers do not examine whether or not a particular DS0 channel is active prior to performing alternate routing; however, some multiplexers may have the ability to perform forced or transparent bumping regardless of whether they use fixed tables or dynamically generate alternate paths. Under forced bumping, DS0 channels are immediately reassigned, whereas, under transparent bumping, current voice or data sessions are allowed to complete prior to their bandwidth being reassigned.

Redundancy

Since the failure of a T-carrier multiplexer can result in a large number of voice and data circuits becoming inoperative, redundancy can be viewed as a necessity similar to business insurance. To minimize potential downtime, you can consider dual power supplies as well as redundant common logic and spare voice and data adapter cards. Doing so may minimize downtime in the event of a component failure as many multiplexers are designed to enable technicians to easily replace failed components.

Maximum number of hops and nodes supported

As T-carrier multiplexers are interconnected to form a network, each multiplexer can be considered as a network node. When a DS0 channel is routed through a multiplexer that multiplexer is known as a hop. Thus, the maximum number of hops refers to the maximum number of internodal devices a DS0 channel can traverse to complete an end-to-end connection.

In addition to the maximum number of hops, users must also consider the maximum number of nodes that can be networked together. The maximum number of addressable nodes that can be managed as a single network is normally much greater than the maximum number of hops supported, since the latter is constrained by the delay to voice as DS0 channels are switched and routed through hops.

Diagnostics

Most T-carrier multiplexers provide both local and remote channel loop-back capability to facilitate fault isolation. Some multiplexers have built-in test pattern generation capability which may alleviate the necessity of obtaining additional test equipment for isolating network faults. The reader is referred to Chapter 6 for specific information concerning the testing of digital facilities and the use of built-in and stand-alone test equipment.

Configuration rules

Figure 5.14 illustrates a typical T-carrier multiplexer cabinet layout which is similar to the manner in which a multiplexer would be installed in an industry standard 19-inch rack. In examining multiplexer configuration rules, a variety of constraints may exist to include the number of trunk module cards, voice cards, and data cards that can be installed. Other constraints will include the physical number of channels supported by each card and the type of voice digitization modules that can be obtained. Depending upon end-user requirements, additional expansion shelves may be required to support additional cards. When this occurs, additional power supplies may be required and their cost and space requirements must be considered.

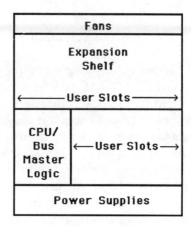

Figure 5.14 Typical T-carrier multiplexer cabinet

5.2.4 Multiplexers and nodal processors

Due to the liberty by which vendors can label products, there is no definitive line that separates a T-carrier terminal or end-unit multiplexer from a nodal processor. In general, we can categorize each device by the number of trunks they support, the method by which alternate routing is performed, and the method by which operators control the network.

In general, a nodal processor that supports more than 16 T-carrier circuits and includes the capability to dynamically perform alternate routing based upon one or more algorithms. In addition, this device normally permits network configuration to be effected from a central node. In comparison, a T-carrier terminal or end-unit multiplexer supports up to 16 trunks and usually relies upon the use of predefined tables to perform alternate routing, assuming they actually have this capability. In addition, network reconfiguration may require operators to reprogram each multiplexer individually.

REVIEW QUESTIONS

1 Why is the data rate on a DS0 channel limited to 56 kbps when the channel is routed through a D4 channel bank?

2 Discuss three differences between channel banks and T-carrier multiplexers.

3 Why would you consider the use of M24 multiplexing or an equivalent service?

4 What is the major benefit obtained from the use of M44 multiplexing or an equivalent service? What is one limitation associated with the use of this service?

5 What are the primary differences between the functionality of T-carrier terminal or end-unit multiplexers and T-carrier nodal switches?

6 Illustrate the advantage obtained by the use of a non-contiguous resource allocation feature of a T-carrier multiplexer.

7 What are two problems associated with the use of applying packet technology to route calls through a series of T-carrier multiplexers?

8 Under what circumstances would you want to use a two-wire or four-wire transmission-only voice interface on a T-carrier multiplexer?

9 What is the maximum number of voice channels that can be carried on T1 and CEPT 30 facilities if the T-carrier multiplexer uses PCM to digitize voice? How do those numbers change when 32 kbps and 24 kbps ADPCM are used?

10 How could you maximize the efficiency of a T-carrier multiplexer that does not support subrate multiplexing?

11 Discuss three functions a gateway T-carrier multiplexer must perform.

12 Describe two methods used by T-carriers to effect the alternate routing of DS0 channels. Which method would normally be more efficient?

13 What does the term downspeed mean?

14 What does the term transparent bumping refer to?

15 What is the difference between a multiplexer functioning as a hop and a multiplexer functioning as a node?

6

TESTING AND TROUBLESHOOTING

While digital transmission facilities can be expected to provide a high level of availability and reliability, they can also be expected to be less than problem free. Due to this, the primary objective of this chapter is to focus attention upon the methods by which digital transmission facilities can be tested and how different types of problems can be isolated, a procedure commonly referred to as troubleshooting. To accomplish this objective we will first examine a variety of performance measurements associated with the use of digital facilities. This will be followed by an investigation into the operation and utilization of several types of test equipment as well as the use of equipment indicators for isolation problems. Finally, due to the growth in the utilization of T1 circuits, we will conclude this chapter by reviewing the electrical specifications and service objectives of that facility that can be used as a basis for testing and troubleshooting.

6.1 PERFORMANCE MEASUREMENTS AND OBJECTIVES

One of the key differences between analog and digital transmission facilities is in the area of performance measurements and objectives. Analog line quality is normally expressed by such terms as signal-to-noise ratio, dB levels, phase (delay) distortion and frequency response. In comparison, the quality of a digital line is normally expressed in terms of the number of errors occurring per unit of time or by the number of units of time in which one or more errors occurred.

Many communications carriers design their digital facilities to provide a level of service in terms of percent error free seconds, number of severely errored seconds, error seconds in an 8-hour day, bit error rate, availability or unavailability percentage. In this section we will first examine the performance measurements associated with digital facilities and then investigate the performance objectives associated with the use of those facilities.

Table 6.1 lists commonly used performance measurements and alarms associated with the use of digital transmission facilities. In the top portion of the table, performance measurements related to a common measurement are so indicated by being indented under a specific measurement.

Table 6.1   Performance measurements and alarms.

Performance measurements
  Bipolar violation rate
  Bit errors
    Bit error rate
  CRC errors
  Delay time
  Errored seconds
    Error-free seconds
    Percent error-free seconds
    Unavailable seconds
    Severely errored seconds
    Consecutive severely errored seconds
    Degraded minutes
  Framing errors
    Severely errored frames
  Pattern slips

Alarms
  Loss of clock
  Loss of synchronization
  Yellow alarm
  Alarm indication signal

6.1.1 Performance measurements

Although Table 6.1 lists 15 common performance measurements, in many situations only a small subset of those measurements may be of concern due to the method by which some communications carriers denote performance on their facilities. By first examining each of the performance measurements listed in Table 6.1, we will obtain a basic understanding of common measurements that will be applicable to most digital facilities.

Bipolar violation rate

A bipolar violation (BPV) occurs whenever two successive ones have the same polarity. Although the bipolar violation rate can indicate a degree of poor line quality or the presence of defective repeaters, this rate can also be misleading. To understand the latter, consider how several coding formats previously discussed in this book are used to provide a minimum ones density through the use of intentional bipolar violations. Thus, using test equipment to monitor BPVs, when such equipment counts both intentional and unintentional bipolar violations, can result in the equipment measuring a correct BPV rate that has no relation to line quality nor the operational status of repeaters. A second problem associated with bipolar violations is the fact that equipment, including multiplexers, CSUs, and digital switches, removes bipolar violations. Thus, a BPV rate is only applicable and useful as a measurement for one section of a digital facility.

Under AT&T's Accunet T1.5 service, an excessive bipolar violation condition is considered to occur when any digital circuit experiences a number of bipolar violations that results in a performance level below a threshold of a 10^{-6} BPV rate for 1000 consecutive seconds. This error rate corresponds to 1544 BPVs in 1000 consecutive seconds, since the line rate is 1.544 Mbps.

An excessive bipolar violation condition may indicate a problem with end-user CSUs or a line or repeater problem that should be reported to the carrier for corrective action. To isolate the cause of the problem to end-user or carrier facilities, end-users can consider operating the self-testing feature built into many CSUs. This self-testing feature causes the internal circuitry of the device to be tested and , if successful, indicates that the problem is elsewhere. Thus, upon the successful self-testing of CSUs at both ends of a T-carrier, an excessive bipolar violation rate should be reported to the communications carrier.

Bit errors

A bit error represents the change of a 'zero' bit to a 'one' bit or a 'one' bit to a 'zero' bit. By counting bit errors and computing a bit error rate based upon the total number of bits transmitted or the total number of bits transmitted during a predefined period of time, one can obtain a bit error rate. Thus, the bit error rate (BER) can be expressed as

$$BER = \frac{\text{bits in error}}{\text{total number of bits transmitted}}$$

The BER is used by several communications carriers as a measure for judging circuit quality and availability based upon a CCITT recommendation discussed later in this chapter.

Table 6.2 Bit error rates.

Bit error rate equivalents

10^{-3}	1 error in 1 000 bits
10^{-4}	1 error in 10 000 bits
10^{-5}	1 error in 100 000 bits
10^{-6}	1 error in 1 000 000 bits
10^{-7}	1 error in 10 000 000 bits
10^{-8}	1 error in 100 000 000 bits
10^{-9}	1 error in 1 000 000 000 bits

T-carrier bit errors

	Bit errors	
Error rate	T1 (1.544 Mbps)	CEPT PCM-30 (2.048 Mbps)
10^{-9}	1 error per 10.79 minutes	1 error per 8.14 minutes
10^{-8}	1 error per 65 seconds	1 error per 48.8 seconds
10^{-7}	1 error per 6.5 seconds	1 error per 4.88 seconds
10^{-6}	1.544 errors per second	2.048 errors per second
10^{-5}	15.44 errors per second	20.48 errors per second
10^{-4}	154.4 errors per second	204.8 errors per second
10^{-3}	1544 errors per second	2048 errors per second

In Table 6.2 the reader will find a list of bit error rates. The top portion of the table indicates the bit error rate equivalents for bit error rates expressed as a power of 10. In the lower portion of

Table 6.2, T-carrier bit errors were computed for North American and European circuits operating at 1.544 Mbps and 2.048 Mbps respectively.

CRC errors

CRC error checking is applicable to extended superframe (ESF) and European PCM-30 T-carrier systems using an optional CRC-4 checking algorithm. Under both framing formats, cyclic redundancy checking (CRC) is calculated on the transmitted data and inserted into a portion of the framing information. When the data checked by the CRC algorithm is received, the CRC is recalculated and compared to the original CRC. If the two CRCs match, the transmitted data is considered to have been received without error. If the two CRCs do not match, one or more bits in the frame covered by the CRC check are considered to have been received in error.

A CRC error count provides a mechanism for counting frame errors. Although it does not provide a precise measurement of error activity as bit errors do, CRC errors can be counted without disturbing the flow of data. In comparison, bit error testing requires injecting a known signal onto a line or channel and comparing the received data to the same bit generation process used to generate the bit sequence. This type of testing interrupts the flow of traffic and is known as intrusive testing.

Delay time

The routing of data through an extensive digital network can result in cumulative delays. Such delays include the time required to switch channels or packets at nodes, T-carrier multiplexer processing time, and propagation delay time. As delay time increases, its effect upon voice and data transmission become significantly different.

If the delay time exceeds approximately 125 to 250 milliseconds, a voice conversation will start to appear awkward, although a listener will still be able to ascertain what was said. If a data transmission session is occurring, an increasing delay time can become the governing factor resulting in protocol time-outs. In such situations the protocol procedure requires a response to each transmitted block of information within a predefined time period. If this response is not received, the protocol may be configured to drop the session, resulting in a failure to communicate over an operational circuit due to an excessive delay time.

Framing errors

A framing error is a logical error that occurs within the framing bits of a digital signal. Under ESF framing, the occurrences where two or more frame errors transpire within a 3-millisecond multiframe are known as a severely errored frame.

Some test sets can be used to count frame errors. Doing so will provide an indication of a bit error rate per 193 or 256 bits if you assume bit errors are randomly distributed. Of more importance, a framing error count will provide an indication of how frequently T-carrier equipment loses synchronization with one another.

Errored seconds

An errored second (ES) is a second during which one or more bit errors occurred. This measurement forms the basis for several other types of related performance measurements to include error-free seconds, severely errored seconds, consecutive severely errored seconds, unavailable seconds, and degraded minutes.

An error-free second (EFS) is a 1-second period of time in which no bit errors occurred. The percentage of error-free seconds then becomes

$$\text{error free seconds (\%)} = 100\% - \frac{\text{error seconds}}{\text{total seconds}} \times 100$$

The percentage of error-free seconds is commonly used as a measurement for qualifying digital circuits at installation as well as providing a basis for the measurement of circuit quality.

A severely errored second (SES) is considered to be any second with a bit error rate greater than or equal to 1×10^{-3}. When one is monitoring CRC errors where the CRC errors represent a frame or block error rate, a severely errored second then represents any second with 320 or more CRC-6 errors. When two or more severely errored seconds occur consecutively, this situation is known as a consecutively severely errored seconds (CSES) event.

Two additional performance measurements related to errored seconds are unavailable seconds (UA) and degraded minutes (DM). An unavailable second is 1 second in a period of time in which ten or more consecutive severely errored seconds occurred. Under CCITT recommendation G.821, which is discussed later in this chapter, the occurrence of this event signifies a degree of line deterioration such that a circuit is considered unavailable for use. A degraded minute is a period of 60 consecutive seconds in which the bit error rate exceeded one error per million bits (1×10^{-6}).

Pattern slips

A pattern slip is the addition or deletion of a bit to a transmitted pattern, such as framing. Pattern slips typically indicate a problem with system timing or repeaters on a digital span.

6.1.2 Alarms

In comparison to performance measurements that provide knowledge about the quality level of facilities and equipment, alarms denote the occurrence of abnormal conditions whose cause must be rectified if communications are to be restored.

Types of alarms

A loss of clock occurs when a device has received a long string of consecutive zeros, while a loss of synchronization occurs when two or more of five framing bits are received in error. Both the yellow alarm and alarm indication signal (blue alarm) are indicated by predefined bit patterns previously covered in Chapter 5.

Alarm simulation

To insure equipment operates correctly, it is often a good idea to use a digital test set as an alarm simulation device. Doing so enables you to observe the response of equipment to simulated alarm conditions and verify if the equipment responds correctly.

6.2 PERFORMANCE CLASSIFICATIONS

The CCITT G.821 recommendation defines four error rate performance categories. These four categories include available and acceptable, available but degraded, available but unacceptable, and unavailable, and are graphically illustrated in Figure 6.1.

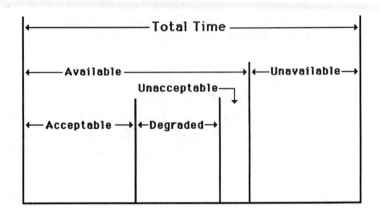

Figure 6.1 Performance classifications

6.2.1 Availability levels

The available and acceptable performance category is based upon intervals of test time of at least 1 minute during which the bit error rate is under 10^{-6}. This performance category results in a good voice quality on a T-carrier facility.

The available but degraded performance category is based upon intervals of test time of at least 1 minute during which the error rate is between 10^{-3} and 10^{-6}. The occurrence of this performance category usually results from microwave fading, atmospheric disturbances, or degraded repeaters.

The available but unacceptable performance category is based upon intervals of test time of at least 1 second, but less than 10 consecutive seconds, during which the error rate is greater than 10^{-3}. On a T-carrier facility, an unacceptable performance level normally results from error bursts, clock slips, and circuit switching hits.

The last performance category defined by the CCITT G.821 recommendation is unavailable. This performance category is based upon intervals of test time of at least 10 consecutive seconds during which the error rate is greater than 10^{-3}. When this situation occurs, a T-carrier facility is considered inoperative and not available for use.

6.2.2 Computing availability

Over a period of time, circuit availability can be computed and compared to a carrier's guaranteed level of performance, which is normally expressed as a percentage of availability. Since available circuit time is total time less unavailable time, availability expressed as a percentage becomes

$$\text{availability (\%)} = \frac{\text{availability}}{\text{available} + \text{unavailable}} \times 100$$

where

$$\text{available} + \text{unavailable} = \text{total time}$$

Although a high level of availability is very desirable, by itself the figure can be misleading. As an example of this, consider Table 6.3 which lists outages you can expect based upon five levels of availability. Note that total time in computing availability is based upon a 24-hour day, even though most organizations may use a digital facility only a fraction of that time. Since many T-carrier facilities are routed over microwave systems, you can normally expect a higher level of electromagnetic interference to occur during daytime. This, in turn, will more than likely result in a lower level of availability during daytime than other periods of the day.

Table 6.3 Availability versus outages.

Percent availability	Monthly outage	Annual outage
99.5	3.7 hours	44 hours
99.8	1.5 hours	17.5 hours
99.9	44 minutes	8.8 hours
99.99	4.4 minutes	53 minutes
99.999	27 seconds	5.3 minutes

Currently, AT&T specifies a 99.9 percent level of availability for its subrate Dataphone Digital Service facilities. For that vendor's Accunet T1.5 1.544 Mbps transmission facility, a 99.7 percent level of availability is specified.

6.3 COMMON TESTS AND TEST EQUIPMENT

In this section we will examine several common tests and the use of test equipment to obtain an indication of the quality of a digital facility as well as the operation of equipment connected to such facilities.

6.3.1 BERT

Bit error rate testing (BERT) involves generating a known data sequence into a transmission device and examining the received sequence at the same device or at a remote device for errors.

Normally, BERT testing capability is built into another device, such as a protocol analyzer or multiplexer. The use of a BERT results in the computation of a bit error rate (BER), which is

$$BER = \frac{\text{bits received in error}}{\text{number of bits transmitted}}$$

To perform a bit error rate test using one tester, communications equipment must be placed into a local loop-back mode of operation and the BERT test can then be used to determine if equipment is operating correctly. If a BERT test is conducted end-to-end via a loop-back at the distant end of the circuit, the test can indicate the level of performance of equipment connected to the facility and the facility. In certain situations equipment at the distant end of a circuit can be placed into a loopback mode of operation in which the digital facilities transmit and receive wires are bridged. When this occurs and a bit error rate tester is directly connected to a DSU or CSU, the resulting measurement can be used as an indication of circuit quality.

Figure 6.2 illustrates how the distant end of a T-carrier circuit can be placed into a loop-back mode of operation. In this example the loop-back can occur at the network interface (NI) installed at the carrier, at the channel service unit (CSU), or at the data terminal equipment (DTE). Loop-backs at the NI or CSU can be effected by sending for a period of 5 seconds a repeating pattern as denoted in Table 6.4 for North American T1 facilities. To loop-back the DTE, such as a multiplexer, requires either manual intervention at the remote site or if the DTE is operating under the control of a central facility the transmission of an appropriate instruction to place the device into loopback.

Once a loop-back is effected you can use one BERT to perform a BER which can be used to obtain an indication of end-to-end channel performance. As an example, a bit error rate in excess of 1 per 1000 during 10 consecutive seconds would indicate

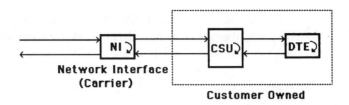

Network Interface
(Carrier)

Customer Owned

Figure 6.2 Loop-back methods. Remote loopbacks can be effected at the network interface, CSU, or at the DTE

Table 6.4 T1 loop-back codes.

Device	Operation	Bit pattern	Result
CSU	SET	10000	Places CSU in loop-back
	RESET	100	Cancels CSU loop-back
NI	SET	11000	Places NI in loop-back
	RESET	11100	Cancels NI loop-back

an unavailable level of performance based upon the CCITT G.821 recommendation.

The most popular test signal for T1 troubleshooting is a repeating pattern of $2^{20}-1$ (1 048 575) bits as specified in AT&T publication 62411. Known as a quasi-random signal source (QRSS), it violates minimum density requirements since bits 2 through 49 of the pattern contain 44 zeros and 4 ones, while a 48-bit string requires at least 5 zeros.

When used for testing a DS1 signal the QRSS pattern is used to simulate the activity on individual or all DS0 channels. All bits in each frame with the exception of the framing bit result from the generation of a repeating 1 048 575 bit pattern. Some test sets have an option which forces an output bit to a binary one whenever the next 14 bits in the pattern will be binary zeros. This option is unnecessary when B8ZS coding is used, since zero suppression is then generated by the CSU that complies with the coding format. Since QRSS testing is disruptive to data traffic, CRC analysis and counting of framing bit errors should be considered. Both of these methods provide a non-disruptive method of obtaining performance data.

Table 6.5 indicates the test time required to calculate bit error rates when a QRSS test pattern is inserted into a single DS0 channel on a T1 circuit. Since the time is proportionally reduced as the number of DS0 channels carrying the signal is increased, you can divide the calculation time by the number of channels used to transmit the QRSS sequence if more than one DS0 channel is used.

Table 6.5 Error rate calculation time per DS0.

Error rate	Calculation time
10^{-3}	1.56 seconds
10^{-4}	15.6 seconds
10^{-5}	156 seconds
10^{-6}	1562 seconds
10^{-7}	28.6 minutes
10^{-8}	51.7 minutes
10^{-9}	286 minutes

T-carrier BERT plug-in modules

Several T-carrier BERTs use plug-in modules to support different types of T-carrier facilities. Interface plug-in modules govern the physical interface such as RS232, CCITT V.35, RS-449; the method of framing, including unframed, D4, ESF, or CEPT PCM-30; the method of line code support, such as B7, B8ZS, HDB3; and countable events, including bipolar violations, framing errors, CRC errors, and bit errors.

In its test mode of operation, the BERT simultaneously counts transmitted and received characters, bits in error and may count bipolar violations, framing errors, and CRC errors. If the error type being counted is CRC errors the bit error rate (BER) may represent a block error ratio depending upon the BERT test device used. As an example, some BERTs count the number of errored CRC blocks by considering an extended superframe of 4632 bits to be one block of data. If a bit error occurs within the six CRC bits, the extended superframe is considered to be a block error. Then, the number of errored CRC blocks is divided by the total number of CRC blocks to obtain a block error ratio.

Another interesting aspect of some BERT test sets is the ability to select patterns that generate a violation of ones density requirements. Then the injection of this type of test sequence can be used to verify the correct operation of equipment performing

B8ZS or HDB3 line coding, in effect, performing a stress test on that equipment.

To convert an error counter number in a BERT to a bit error rate, testing for a fixed period of time is required. Table 6.6 lists the time required for two common bit error rates based upon seven distinct data rates. Note that this table can be used for testing low-speed multiplexer channels as well as currently available subrate digital services.

Table 6.6 Bit error rate versus test times.

Data rate	Bit error rate		
(bps)	1×10^{-5}	1×10^{-6}	14×10^{-9}
300	5 min 33 sec	55 min 33 sec	55 555 min
600	2 min 47 sec	27 min 47 sec	27 777 min
1200	1 min 23 sec	13 min 53 sec	13 888 min
2400	42 sec	6 min 57 sec	6 944 min
4800	21 sec	3 min 28 sec	3 472 min
9600	11 sec	1 min 44 sec	1 736 min
19.2K	6 sec	52 sec	868 min
56K		17.8 sec	297 min
64K		15.6 sec	260 min
1.544M			10.8 min
2.048M			8.14 min

As an example of the use of Table 6.6, consider the 9600 bps data rate for testing purposes. If, during a test time of 11 seconds, exactly 3 bit errors occurred, then the bit error rate is 3×10^{-5}. To illustrate the conversion of one bit error rate to another bit error rate assume a BERT was performed on a PCM-30 system operating at 2.048 Mbps for precisely 8.14 minutes and the bit error counter displays 1631 bits in error. This would indicate a BER of 1631×10^{-9}, or 1.631×10^{-6}.

Since most protocols group data into blocks for transmission, a block error rate may provide a more realistic level of performance than a bit error rate. This is because a burst of noise resulting in a high bit error rate may affect fewer blocks and result in a lower block error rate than a lower bit error rate where errors are more evenly distributed over time.

Error rate testing methods

Figure 6.3 illustrates three common methods whereby bit error rate testing can be used on digital facilities. In each example the unlabeled rectangles represent either a combined DSU/CSU or a CSU, with the type of device dependent upon whether subrate or T-carrier facilities are being tested.

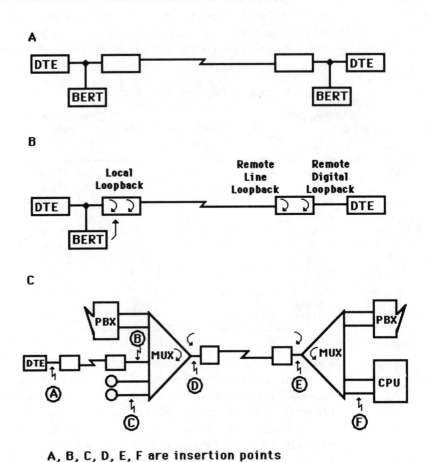

A, B, C, D, E, F are insertion points

Figure 6.3 BERT testing: A end-to-end; B fault isolation using loop-backs; C network component testing

The end-to-end testing illustrated in Figure 6.3A that is performed without a loop-back requires the use of two BERTs. The advantage in using this method of testing is that problems can be isolated to either side of the transmission line.

In Figure 6.3B, the process of performing fault isolation using four types of loop-backs is illustrated. In comparing parts A and B of Figure 6.3, note that through the use of loop-backs you can perform testing with one test set instead of two. In addition, through the use of loop-backs you obtain the ability to test the operational status of both selected equipment and portions of the circuit. To illustrate this let us examine the four loop-backs illustrated in Figure 6.3B.

The first loop-back moving from left to right is commonly called a local digital loop-back. This loop-back ties transmitter to receiver without data being modulated into a bipolar signal and then demodulated back into a unipolar signal. Using this loop-back, in effect, echoes data back to the DTE or test instrument. If a DTE was used by itself when a local digital loop-back was in effect, transmitted characters would be echoed back to the device. Thus, this loopback can be used to test both the operation of the DTE as well as the cable connecting the DTE to the DSU or CSU. Similarly, using a BERT would test the cable connection to the DSU or CSU.

The second loop-back is commonly referred to as a local analog loop even though it results in the DSU/CSU or CSU converting unipolar digital signals to bipolar digital signals and back to unipolar. The reason for the term local analog loop-back is that it references the function modems perform when placed in that type of loop-back. Although DSU/CSU or CSUs modulate digital data, the term continues to be used as a hold-over from analog terminology. When placed in a local analog loop-back mode of operation, the BERT tests the complete DSU/CSU or CSU, including its modulation and demodulation circuitry. Note that neither the local digital nor the local analog loopbacks permit the high-speed circuit to be examined.

The third loop-back, commonly referred to as a remote line loop-back or remote digital loop-back, connects the transmit and receive wires of the circuit. This functions as a digital bridge, passing bipolar violations back to the local site without correction. Thus, a test instrument inserted at the local site on the receive side of the line would obtain an accurate count of bipolar violations on the loop when the remote DSU/CSU or CSU is placed into a line loop-back mode of operation.

The fourth loop-back is similar to the second loop-back with respect to the fact that digital data in the line's bipolar format is converted to unipolar data and then back to bipolar for transmission back to the local site. This loop-back loops in the digital section of the DSU/CSU or CSU and converts bipolar violations to binary logic errors prior to encoding the resulting unipolar signal back to a bipolar signal for transmission.

Although the remote line loop-back provides a mechanism to measure the BPV rate, it also provides a mechanism to obtain the BER on the line since a simple bridge operation is performed by the remote device to tie the circuit's transmit and receive wires together. When a remote digital loop-back is effected, a test from the local site not only measures line quality but, in addition, the operational status of the remote DSU/CSU or CSU.

Both local analog and local digital loop-backs are normally set by the use of switches or buttons on the local device. To effect remote loop-backs, some devices, including locally installed DSU/CSUs and CSUs, will generate appropriate loop codes when switches or buttons labeled 'Remote' are toggled or pressed. Typically, these buttons or switches have two position settings labeled set and reset, with placement to the set position causing an activate loop-back code to be transmitted to the remote device.

Insertion points

When communications equipment to include multiplexers is connected via digital facilities, both the number of insertion points that can be used for testing, as well as the types of loop-backs one can use for testing, can substantially increase. This is illustrated in Figure 6.3C. Here the curved lines inside each multiplexer indicate the loop-back of a specific channel or channels for testing of lines connected to each multiplexer. The curved lines outside each multiplexer indicate a remote loop-back which can be performed on a channel basis or on the entire carrier circuit. Finally, the circled letters represent a few of the possible insertion points where test equipment can be used in conjunction with the loop-back of a specific multiplexer channel to ascertain the performance of equipment or facilities similar to the methods described previously.

6.3.2 Error-free second tester

In an error-free second (EFS) test, received data is analyzed on a per second basis. If one or more bit errors occurs during a 1-second interval, the interval is recorded as an errored second.

An error-free second tester can be employed in a manner similar to that described for bit error rate testing. That is, the tester can be used with four types of loop-backs described for DSU/CSU or CSU operations or the tester can be employed using loop-backs that are available through the use of different types of DTEs.

EFS testers incorporate a variety of interface connectors that can include V.35 (34 pin), RS-449 (37 pin), RS-232 (25 pin), or X.21 (15 pin) interface connectors for use on digital facilities. Typical data rates supported by ESF testers include subrates from 2.4 kbps to

56 or 64 kbps, as well as T-carrier 1.544 and 2.048 Mbps operating rates.

As previously discussed in this chapter, many communications carriers specify the availability of their high-speed digital facilities in terms of errored seconds or error-free seconds over a specified time period. Table 6.7 compares the CCITT G.821 error performance recommendation with British Telecom's 64 kbps digital service goal, and AT&T DDS and Accunet T1.5 service objectives with the latter for circuits with distances exceeding 1000 miles. In actuality, the performance objectives of AT&T's Accunet T1.5 circuits are complex and are based upon both the type of line segment as well as the length of the segment. For customer premises-to-customer premises distances less than 250 miles, the error-free seconds performance goal is 96.6 percent. For a circuit length from 250 to 1000 miles the goal is 96.0 percent, while distances greater than 1000 miles have a goal of 95.0 percent error-free seconds. For serving office-to-serving office circuit segments the performance objectives are 99.1 percent, 98.5 percent, and 97.5 percent error-free seconds for distances less than 250 miles, distances from 250 to 1000 miles, and distances greater than 1000 miles respectively. For local exchange carrier provided access from the customer premises to the serving office the performance objective is 98.75 percent error-free seconds.

Table 6.7 Digital circuit error performance.

	% Error-free seconds	Error in 8-hour day
CCITT recommendation G.821 for 64 kbps service	98.8	346
British Telecom 64 kbps KiloStream/MegaStream goal	99.5	144
Dataphone Digital Service	99.5	144
Accunet T1.5	95.0	1 440

6.3.3 T-carrier simulator

A T-carrier simulator permits users to predict how a system will perform. The simulator permits users to corrupt the T-carrier framing pattern to determine how a device responds to framing errors, framing errors at a rate less than that needed to cause a loss of frame synchronization, or framing errors at a rate equal or above that necessary to cause a loss of synchronization.

Through the use of a T-carrier simulator, you can determine if yellow and red alarms are generated and if a device under test responds correctly, as well as determine the time required to recover from a loss of synchronization. Other practical uses of a T-carrier simulator include modeling data link impairments under controlled programmable conditions, testing the development of hardware and software off-line prior to their actual installation and operation, and the periodic performance testing of hardware and software.

6.3.4 T-carrier transmission test set

A T-carrier transmission test set is the most sophisticated of all types of test equipment that can be used with digital facilities. Equipment in this category can be used to perform frame level tests, gather statistics concerning different types of error conditions, and may include the ability to remove a channel from a T-carrier line and apply traditional analog or digital testing to that channel. Table 6.8 lists some of the typical measurements T-carrier transmission test sets are capable of performing. Since most of these performance measurements were previously discussed in this chapter, let us focus our attention upon those measurements not previously discussed as well as certain elements of specific measurement that warrant an elaboration.

Jitter and wander

Both jitter and wander can be considered as a potential disaster waiting to occur. To understand the basis for each impairment it is necessary to first understand how timing delays occur in a digital network.

As data pulses flow through a digital network, repeaters recover pulse clocking from the incoming data. Unfortunately, this recovery is not instantaneous as there are built-in delays in the circuitry which recognize an incoming signal and then regenerate the signal. Another source of timing delays are digital multiplexers that build subrate facilities onto a T-carrier circuit. Such multiplexers add bits, a process called bit stuffing, to synchronize the low-speed incoming digital pulses to the T-carrier's operating rate. Since bit insertion will not occur at a precise time due to circuitry delays, the multiplexer is another source of timing delays.

Table 6.8 T-carrier transmission test set measurements.

Frame errors and alarms
    Frame bit error ratio
    Severely errored seconds
    Out of frame seconds
    Loss of frame (red alarm)
    Loss of frame seconds
    Yellow alarm seconds

CRC-6 errors (ESF) and CRC-4 errors (CEPT PCM-30)
    Average and current BER
    Errored seconds
    % error-free seconds

Jitter and wander
    Jitter hit seconds
    Phase hit seconds

Line and signal measurements
    BPV count
    BPV errored seconds
    Excess zeros

Simulation
    All ones (blue alarm)
    Red alarm
    Yellow alarm

The difference between the ideal and the actual time of arrival of a digital pulse is known as jitter and wander, with the term used dependent upon the magnitude of the difference. If the difference is at a rate less than 10 Hz it is known as wander, while a difference between the ideal and actual arrival time of a pulse that exceeds 10 Hz is known as jitter. Since both wander and jitter are cumulative they will eventually build up to a point where network synchronization will be lost, resulting in a condition in which a bit time is eventually either gained or lost as illustrated in Figure 6.4.

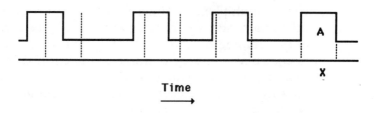

Time

Figure 6.4 Displacement in time of a signal. Although bit A at time X is in its exact clock position, the cumulative jitter will result in succeeding bits being displaced more and more by time

To compensate for the effect of wander and jitter, digital network elements contain buffers, functioning as a compensation unit between the clocking differences building up in the network. As an example, data entered into a multiplexer at one rate is first buffered prior to transmission at a different rate. Thus, buffers in effect can remove accumulated jitter as long as the variations do not exceed their capacity. When an excessive amount of jitter or wander occurs, the buffers in network equipment may either overflow or underflow. When an overflow condition occurs due to data pulses arriving early, the buffer must delete a block of data to maintain synchronization. In comparison, when data pulses arrive late, this condition results in an underflow in which the buffer must repeat a block of data to maintain synchronization with the T-carrier operating rate. Both of these conditions are known as a slip which can be defined as the occurrence of a digital signal buffer overflow or underflow.

When transmission occurs on T-carrier facilities, digital network equipment will intentionally delete or repeat all or a portion of the bits in a frame. When the network equipment performs a slip by deleting or repeating all frame bits, this process is known as a controlled slip. When the network equipment either deletes or repeats a portion of a T-carrier frame, the process is called an uncontrolled slip.

Controlled slips are performed by T-carrier equipment that peforms switching and cross connection functions, such as DACS, PBXs, and central office switching systems. Uncontrolled slips result from the underflow or overflow of buffers that are smaller than the size of a frame, hence they are referred to as unframed buffer slips.

Unframed buffers are incorporated into higher rate multiplexers that are not synchronized to a common frequency source, which, in effect multiplex data sources asynchronously. Examples of equipment with unframed buffers include AT&T M13 and M23 multiplexers as well as the dejitterizing circuitry in end-user T-carrier multiplexers. Unfortunately, an uncontrolled slip can be much more serious than a controlled slip as it results in the shift of the framing bit positions. This shift of the framing bit position is known as a change of frame alignment (COFA) which will result in an out of frame condition at the receiving T-carrier multiplexer, causing the multiplexers to resynchronize for a period of time in which data cannot be passed.

CCITT recommendation G.822 specifies service objectives for international 64 kbps digital facilities. Under this recommendation there should be five or less slips during a 24-hour period 98.9

percent of the time, more than five slips per 24-hour period but no more than 30 slips in any hour less than 1 percent of the time, and more than 30 slips per hour less than 0.1 percent of the time.

Line and signal measurements

To effectively monitor bipolar violations the test set must not count intentional violations as errors, such as B8ZS or HDB3 coding. In addition, care should be taken as to the placement of the test set since CSUs are designed to remove bipolar violations.

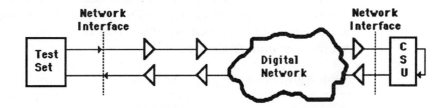

Figure 6.5 Troubleshooting problems at the network interface

Figure 6.5 illustrates the correct insertion of a test set into a digital facility to count BPVs and BPV errored seconds. Note that the test set must have a built-in CSU to transmit bipolar data onto the network at the network interface as well as the ability to generate a remote CSU loop-back code to place the remote CSU into loop-back.

By measuring BPV data you can determine the performance of digital repeaters as well as obtain information concerning noise and crosstalk on the facility since those impairments also contribute toward generating BPVs. Since digital radio, satellites, and fiber optics do not use a bipolar transmission format, this test, in many instances, indicates premises-to-central office performance and is not a measure of end-to-end performance.

6.4 T1 ELECTRICAL SPECIFICATIONS AND SERVICE OBJECTIVES

One of the key problems associated with the testing of a T1 circuit is in attempting to determine if the measured and observed circuit characteristics indicate a problem. In this section we will review T1 electrical specifications and service objectives that can be used by readers as guidelines for testing T1 circuits.

6.4.1 Electrical specifications

In general, the electrical specifications of a T1 circuit define the pulse characteristics of the signal, including its pulse type, pulse rate, pulse shape, and pulse density. Concerning pulse type, a T1 circuit uses bipolar return to zero signaling with 0 volts as the base reference. The pulse rate is 1.544 Mbps; however, many carriers permit a deviation of plus/minus 75 bps. Similarly, the 3 volt dc pulse height or amplitude can vary by plus/minus 10 percent, while the pulse width of 324 nanoseconds can normally vary by plus/minus 45 ns. Lastly, as previously noted, T1 terminal equipment cannot transmit more than 15 consecutive zeros when B7 zero code suppression is used or a bipolar violation must occur when a byte is all zero when B8ZS coding is supported. Table 6.9 summarizes the electrical specifications of a T1 carrier that can be used for testing guidelines.

Table 6.9 T1 electrical specifications.

Specification	Expected value
Pulse type	Bipolar return to zero
Pulse rate	1.544 Mbps +/− 75 bps
Pulse amplitude	3 V dc +/− 10%
Pulse width	324 +/− 45 ns
Pulse density	< 15 consecutive zeros B7 zero code suppression 1 bipolar violation per zero byte B8ZS coding

6.4.2 Service objectives

The service objectives of a T1 facility are expressed by carriers in a variety of ways, including percent error-free seconds, bit error rate, and availability. In general, most carriers have a goal to provide T1 users with a 99 percent or above level of availability per month and an error-free second (EFS) rate at or above 95 percent over each 24-hour period. While most carriers focus upon EFS and availability, a bit error rate exceeding 1 in 1 000 000 is normally considered poor for this type of facility and deserves reporting even if the carrier does not express service objectives in terms of a bit error rate.

By examining the electrical specifications and service objectives of your facility and comparing observed values to those previously discussed in this section, you may be able to quickly isolate problems. If testing does not indicate any deviations from the previously indicated values, you may then want to consider performing other tests and parameter measurements previously described in this chapter and compare those results against the detailed specifications provided by the communications carrier.

REVIEW QUESTIONS

1 How could an excessive bipolar violation rate be a false indication of the quality of a digital circuit?

2 What types of problems could an excessive bipolar violation rate indicate? How could you isolate the problem area to end-user equipment or communications carrier facilities?

3 If the expected error rate on a CEPT-30 facility is 10^{-7}, how many errors can you expect to occur during a 1-minute period?

4 Discuss an advantage and disadvantage of CRC errors with respect to bit errors.

5 What is intrusive testing?

6 Describe the effect of an increasing delay time upon voice and data transmission.

7 What is an error free second?

8 What is the relationship between error-free seconds, error seconds, and total seconds?

9 What is a severely errored second? What is the relationship between a severely errored second and consecutive severely errored seconds?

10 What is an unavailable second?

11 What is a pattern slip and what does its occurrence most likely indicate?

12 Why would you consider using a test set to simulate an alarm condition to a T-carrier multiplexer?

13 Explain why a high level of availability cited by a communications carrier may not be experienced by an organization whose transmission requirements primarily occur between 6 am and 5 pm.

14 What are the three types of loop-backs that can occur on a T1 circuit and how is each loop-back effected?

15 Assume you conducted a bit error rate test on a 128 kbps portion of a T1 circuit. If 88 bit errors occurred during a 51.7 minute period, what is the bit error rate per DS0 channel?

16 Why would you perform a local digital loop-back prior to conducting a BERT test?

17 When should you consider using a T1 simulator?

18 What are two sources of timing delays on digital networks?

INDEX

Index compiled by Paul Nash